SpringerBriefs in Computer Science

More information about this series at http://www.springer.com/series/10028

Niklas Büscher • Stefan Katzenbeisser

Compilation for Secure Multi-party Computation

Niklas Büscher
Security Engineering Group
Technische Universität Darmstadt
Darmstadt, Germany

Stefan Katzenbeisser
Security Engineering Group
Technische Universität Darmstadt
Darmstadt, Germany

ISSN 2191-5768 ISSN 2191-5776 (electronic)
SpringerBriefs in Computer Science
ISBN 978-3-319-67521-3 ISBN 978-3-319-67522-0 (eBook)
https://doi.org/10.1007/978-3-319-67522-0

Library of Congress Control Number: 2017954354

Printed on acid-free paper

This Springer imprint is published by Springer Nature
The registered company is Springer International Publishing AG
The registered company address is: Gewerbestrasse 11, 6330 Cham, Switzerland

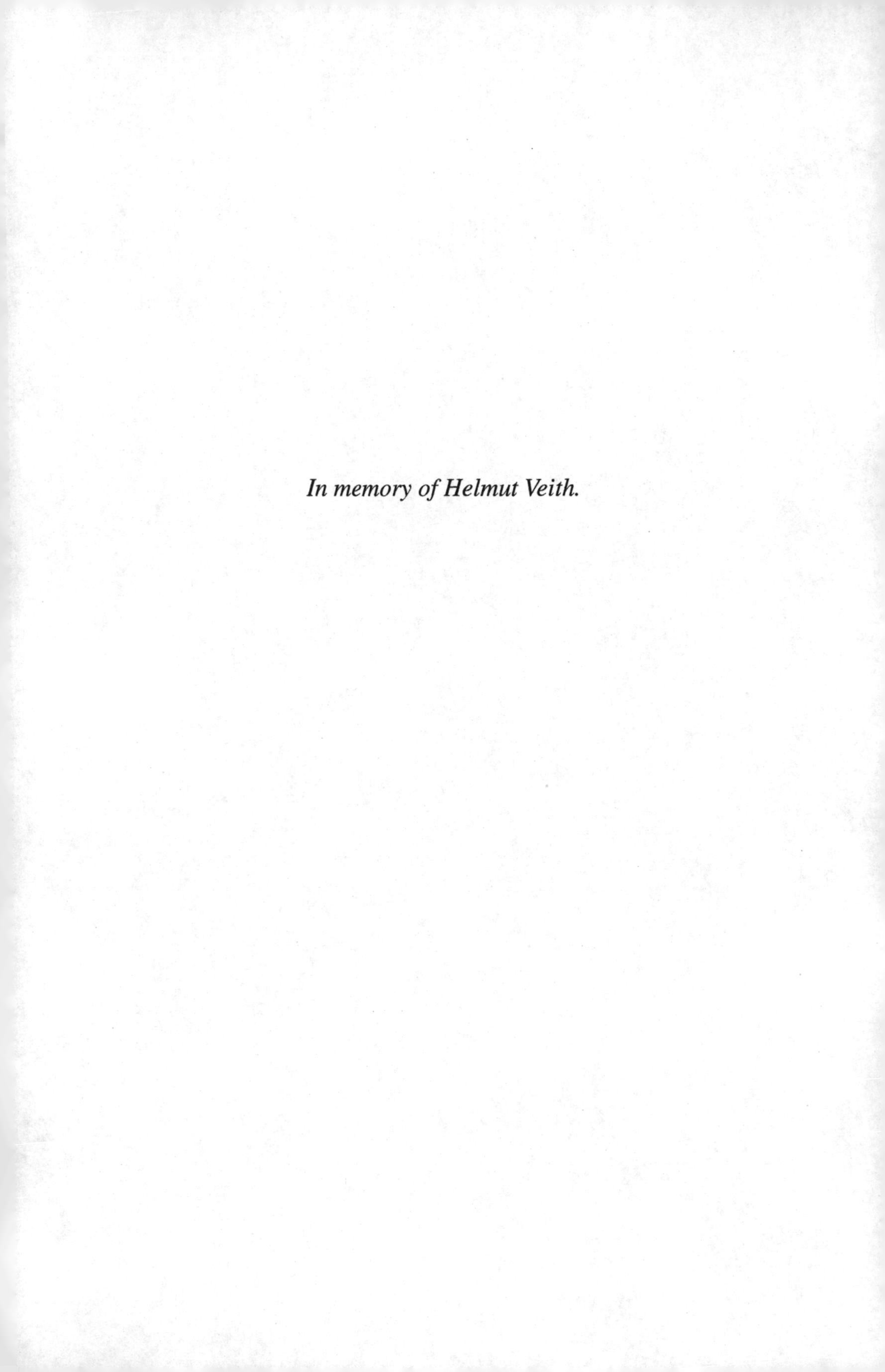

In memory of Helmut Veith.

Preface

An ever-growing amount of data is processed by online services every day. Unfortunately, data owners are at risk to lose control over their possibly sensitive data once it is deployed at an untrusted third party. A solution to this dilemma is Secure Multi-party Computation (MPC), which has emerged as a generic approach to realize Privacy-Enhancing Technology in a cryptographically sound manner, as it allows to perform arbitrary computations between two or more parties over encrypted data. In recent years, numerous protocols and optimizations have made MPC practical for relevant real world scenarios. This development has led to a significant improvement in the performance and size of applications that can be realized with MPC.

A major obstacle in the past was to generate MPC applications by hand. Recently special compilers have been developed to build all kinds of applications. In this book, we summarize our research on the compiler CBMC-GC for MPC over Boolean circuits. We show and explain how efficient MPC applications can be created automatically from ANSI-C, which bridges the areas of cryptography, compilation and hardware synthesis. Moreover, we give an insight into the requirements for creating efficient applications for MPC, and thus we hope that this work can be of interest not only to researchers in the area of MPC but also developers realizing practical applications with MPC.

The authors wish to thank Andreas Holzer, Martin Franz and Helmut Veith, with whom we started the research on compilers for MPC. Moreover, we wish to thank our students Alina Weber and David Kretzmer who contributed ideas and implementations to the later parts of this book. This work has been co-funded by the DFG as part of project S5 within the CRC 1119 "CROSSING" and by the DFG within the RTG 2050 "Privacy and Trust for Mobile Users".

Darmstadt, Germany
August 2017

Niklas Büscher
Stefan Katzenbeisser

Contents

1 Introduction 1
1.1 Motivation 1
1.2 Classification of MPC Compilers 3
1.3 Outline of the Book 4

2 Background 5
2.1 Boolean Circuits 5
2.2 Secure Computation 6
2.2.1 Oblivious Transfer 7
2.2.2 Yao's Garbled Circuits Protocol 7
2.2.3 Goldreich-Micali-Wigderson (GMW) Protocol 9
2.3 Benchmarking Applications for MPC Compilers 10

3 Compiling ANSI-C Code into Boolean Circuits 15
3.1 Motivation and Overview 15
3.2 Background: Bounded Model Checking 17
3.3 CBMC-GC's Compilation Chain 18
3.3.1 Input Language and Circuit Mapping 18
3.3.2 C Parser and Type Checking 19
3.3.3 GOTO Conversion 20
3.3.4 Loop Unrolling 21
3.3.5 Conversion into Single Static Assignment Form 23
3.3.6 Expression Simplification 24
3.3.7 Circuit Instantiation 24
3.4 Complexity of Operations in MPC 26

4 Compiling Size-Optimized Circuits for Constant-Round MPC Protocols 29
4.1 Motivation and Overview 29
4.2 Circuit Minimization for MPC 32
4.3 Building Blocks for Boolean Circuit Based MPC 33
4.4 Gate-Level Circuit Minimization 35

4.5 Evaluation ... 39
4.5.1 Evaluation of Circuit Minimization Techniques ... 39
4.5.2 Compiler Comparison ... 41

5 Compiling Parallel Circuits ... 43
5.1 Motivation and Overview ... 43
5.2 Parallel Circuit Evaluation ... 44
5.3 Compiler Assisted Parallelization Heuristics ... 46
5.3.1 Fine-Grained Parallelization (FGP) ... 46
5.3.2 Coarse-Grained Parallelization (CGP) ... 48
5.4 Evaluation of Parallelization in Yao's Garbled Circuits ... 52
5.4.1 UltraSFE ... 52
5.4.2 Evaluation Methodology ... 53
5.4.3 Circuit Garbling (Offline) ... 55
5.4.4 Full Protocol (Online) ... 58

6 Compiling Depth-Optimized Circuits for Multi-Round MPC Protocols ... 61
6.1 Motivation and Overview ... 61
6.2 Compilation Chain for Low-Depth Circuits ... 62
6.2.1 Preprocessing Reductions ... 63
6.2.2 Sequential Arithmetics and Carry-Save Networks (CSNs) ... 65
6.2.3 Optimized Building Blocks ... 67
6.2.4 Gate Level Minimization Techniques ... 71
6.3 Experimental Evaluation ... 71
6.3.1 Benchmarked Functionalities and Their Parameters ... 72
6.3.2 Compiler Comparison ... 73
6.3.3 Evaluation of the Optimizations Techniques ... 74
6.3.4 Protocol Runtime ... 75

7 Towards Scalable and Optimizing Compilation for MPC ... 79
7.1 Motivation and Overview ... 79
7.2 Adapted Compilation Chain ... 80
7.2.1 Compilation Architecture ... 80
7.2.2 Global Constant Propagation ... 81
7.2.3 Implementation ... 83
7.3 Experimental Evaluation ... 84
7.3.1 Description of Experiments ... 84
7.3.2 Compilation Results ... 85

A CBMC-GC Manual ... 87

References ... 89

Chapter 1
Introduction

1.1 Motivation

Recent years have seen an increase in applications that collect and analyze private data on potentially untrusted machines. With the predominant use of cloud services, computations that were previously performed on local computing environments tend to be outsourced to service providers. At the same time, the trend towards "big data" yields to unprecedented collections of sensitive private data. As a remedy to negative effects on privacy, Privacy-Enhancing Technologies (PETs) have emerged as prime technical protection mechanism. PETs follow the idea of "privacy by design", demanding that privacy aspects need to be taken into account during the entire engineering process of a product or service.

A key element of many modern PETs are cryptographic techniques which allow to keep sensitive data encrypted. Traditional cryptography only protects data "at rest" or "in transit"; once data was used in computation it needed to be decrypted. Thus, interest in the cryptographic community soon turned towards encryption technologies that support processing operations on ciphertexts, so that arbitrary functions can be computed "under encryption". In a simple outsourcing setting, one party may encrypt several messages and upload the encryptions to a cloud server, which can then compute any function on the encrypted messages, without requiring to access them in the clear; the encrypted result is subsequently be transferred back to submitting party. In a more general setting, several parties come together and wish to compute a function on a set of private inputs, which never get revealed to any other party. In this case, we speak of Secure Multi-Party Computation (MPC) or Secure Two-Party Computation (TPC) for the special case of two parties.

Cryptographic techniques for secure outsourcing, MPC or TPC have been proposed by the cryptographic community since the 1980s and can be founded on various techniques, such as garbled circuits, secret sharing or homomorphic encryption. While initially being deemed of theoretic interest only, MPC and TPC turned into practical constructions, which are at the core of many modern PETs.

N. Büscher, S. Katzenbeisser, *Compilation for Secure Multi-party Computation*, SpringerBriefs in Computer Science, https://doi.org/10.1007/978-3-319-67522-0_1

At the heart of MPC (or TPC) is the ability to efficiently represent the functionality to be computed. Many existing constructions require either a formulation in terms of a Boolean or an Arithmetic circuit. Circuits are acyclic graphs, where data flows along edges ("wires"), and is processed at nodes, representing either Boolean functions (in case of Boolean circuits) or basic Arithmetic operations (in case of Arithmetic circuits), see Sect. 2.1. Historically, the first MPC applications required to hand-code such circuits, i.e., to efficiently formulate the function to be computed in terms of basic Boolean or Arithmetic operations. Soon, it became apparent that this is infeasible even for moderately complex functions, as the manual construction of efficient circuits for MPC is a complex, error-prone, and time-consuming task. Writing efficient MPC applications required deep knowledge of both classical hardware synthesis, as well as applied cryptography. This task gets even more pressing, as MPC is on the verge to handle circuits that are larger than those produced in classic chip design.[1] The fastest designs are currently able to process one billion gates per second.

Therefore, when introducing the first practical MPC protocol, Malkhi et al. [57] realized the need for tool support. Since then, multiple compiler prototypes for use in MPC have been proposed, which generate circuits out of a high-level description of the functionality to be computed.

We observe that the compilation of Boolean circuits for MPC shares many similarities with classic hardware synthesis. Yet, they differ in multiple important factors. First, both differ in their application. Circuits for MPC describe interactive (end-user) applications, whereas hardware synthesis commonly aims at developing processing or accelerating units. Second, circuits for MPC protocols purely rely on combinational logic rather than sequential logic, because building blocks for reusable storage, such as flip-flops, are not available.[2] Third, practically no layout or space considerations have to be made when designing circuits for MPC, as they are only evaluated virtually. Fourth, the costs for different types of gates differ significantly. In classic logic synthesis, Boolean NAND gates are favored over XOR gates due to their placement costs. However, as shown in Sect. 2.2.2, in most MPC protocols the evaluation of linear gates is practically for free, whereas the evaluation of non-linear gates is costly. Therefore, classical hardware synthesis tools cannot directly be used for this purpose; special compilers are required that produce minimal circuits according to the cost model of MPC.

[1]For example, the most recent accelerator of NVIDIA, the Tesla P100 Computing Platform, has 15 billion transistors; running a secure computation using garbled circuits on 15 billion gates requires less than 20 min on modern infrastructures.

[2]We note that MPC protocols can be combined with protocols for oblivious storage, e.g., ORAM [38], under various security and performance trade-offs. These constructions are beyond the scope of this work.

1.2 Classification of MPC Compilers

Boolean circuit compilers for MPC can be categorized according their capability to compile from a (minimalistic) *domain-specific language (DSL)* or from a widely used *common programming language*. Moreover, compilers can be *independent* or *integrated* into an MPC framework, which finally performs the cryptographic operations. Integrated compilers produce an intermediate representation, which is *interpreted* (instantiated by a circuit) only during the execution of an MPC protocol. These interpreted circuit descriptions commonly allow a more compact circuit representation. Independent compilers create circuits independent from the executing framework, and thus have the advantage that produced circuits can be optimized to the full extent during compile time and are more versatile in their use in MPC frameworks. Some integrated compilers support the compilation of *mix-mode secure computation*. Mix-mode computation allows to write code that distinguishes between oblivious (private) and public computation. This leads to an even tighter coupling between compiler and execution framework, but allows to express a mix-mode program in a single language. Moreover, some integrated compilers support the compilation of programs for *hybrid* secure computation protocols, which combine different cryptographic approaches.

We begin by giving an overview on compilers that use a DSL as input language. The Fairplay framework by Malkhi et al. [57] started the research on practical MPC. Fairplay compiles a domain specific hardware description language called SFDL into a gate list for use in Yao's Garbled Circuits. Following Fairplay, Henecka et al. [40] presented the TASTY compiler, which allows the combination of Yao's garbled circuits with additively homomorphic encryption and is thus an instance of a hybrid approach. TASTY compiles from its own DSL, called TASTYL.

The PAL compiler by Mood et al. [62] aims at low-memory devices as the compilation target. PAL also compiles Fairplay's hardware description language. The KSS compiler by Kreuter et al. [51] is the first compiler that shows scalability up to a billion gates and uses gate level optimization methods, such as constant propagation and dead-gate elimination. KSS compiles a DSL into a flat circuit format. TinyGarble by Songhori et al. [73] uses (commercial) hardware synthesis tools to compile circuits from hardware description languages such as Verilog or VHDL. On the one hand, this approach allows to use a broad range of existing functionalities in hardware synthesis, but on the other hand also shows the least degree of abstraction, by requiring the developer to have experience in hardware design. We remark that high level synthesis from C is possible, yet, as the authors note, this leads to significantly less efficient circuits. Recently, Mood et al. [61] presented the Frigate compiler, which aims at scalable and extensively tested compilation of another DSL. Frigate and TinyGarble produce compact circuit descriptions that compress sequential circuit structures.

Examples for mix-mode frameworks that compile from a DSL are L1, ObliVM, and Obliv-C. Similar to TASTY, the L1 compiler by Schröpfer et al. [72] compiles a DSL into a mixed-mode protocol including homomorphic encryption. ObliVM by Liu et al. [56] extends SCVM [55], which both compile a DSL that support the combination of oblivious data structures with MPC. This approach allows the efficient development of oblivious algorithms. Yet, both compilers provide only very limited gate and source code optimization methods. Obliv-C by Zahur and Evans [77] also supports oblivious data structures, but follows a different, yet elegant approach by compiling a modified variant of C into an executable application that also supports mix-mode computations.

The CBMC-GC compiler [42] is the first of two compilers that creates circuits for MPC from a common programming language (ANSI-C). CBMC-GC follows the independent compilation approach and produces a single circuit description. Moreover, CBMC-GC applies source code optimization through symbolic execution and utilizes state of the art circuit synthesis minimization methods. This compiler will be the subject of the present book. The PCF compiler by Kreuter et al. [52] also compiles C, using the portable LCC compiler as a frontend. PCF compiles an intermediate bytecode representation given in LCC into an interpreted circuit format. PCF shows greater scalability than CBMC-GC, yet only supports comparably limited optimization methods that are only applied locally for every function.

1.3 Outline of the Book

In this book we focus on the compiler CBMC-GC, which is an independent MPC compiler and is able to compile functionality descriptions expressed in ANSI C syntax into Boolean circuits usable for MPC. The compiler itself tries to produce circuits that are as optimized as possible, and which can be run in any MPC protocol implementation that is based on Boolean circuits. Thus, CBMC-GC does not perform secure computations as such, it basically provides an appropriate formulation of the function to be computed. The compiler can be used either in conjunction with MPC or TPC; still, we explain its use with two-party Yao's garbled circuits.

We begin with an introduction of the basic primitives, i.e., Boolean circuits and secure computation in Chap. 2. We will explain the general approach to transform ANSI C code into Boolean formulas in Chap. 3. A key factor for usable and scalable MPC solutions are optimization techniques, which guarantee small circuits, given the cost model of typical MPC computations. We will demonstrate the circuit minimization approaches applied by CBMC-GC in Chap. 4. The subsequent Chaps. 5 and 6 deal with special aspects of compilation, namely parallelism and depth-optimization. Parallelism is a key concept to reduce latency in MPC applications, while depth-minimal circuits are important if secure computation protocols are utilized that require a number of communication rounds. We show in these chapters that the general architecture of CBMC-GC can be adapted to handle these cases. Moreover, in Chap. 7 we will illustrate how the optimization techniques presented in the previous chapters can be extended to compile very large programs, and thus, show how to construct a scalable and optimizing compiler. Finally, the appendix contains a brief manual of CBMC-GC and demonstrates how the compiler can be used to generate practical MPC solutions.

Chapter 2
Background

2.1 Boolean Circuits

Boolean Circuits Boolean circuits are a common representation of functions in computer science and a mathematical model for digital logic circuits. A Boolean circuit is a directed acyclic graph (DAG) with l inputs, m outputs, and s gates. Technically, the graph consists of $|V| = l + m + s$ nodes and $|E|$ edges, where each node can either be of type input, output or gate. Due to the relation of digital circuit synthesis, the edges are called wires. We note that a Boolean circuit, as defined above, is a combinatorial circuit, which is the supported circuit type for most MPC protocols and is therefore used throughout this book. Combinatorial circuits are different to sequential circuits, which can be cyclic and carry a state.

Boolean Gates Each gate in a Boolean circuit has one or multiple input wires and one output wire, which can be input to many subsequent gates. In this book, we will only consider gates with at most two input wires, namely *unary* and *binary* gates. Each gate in a circuit represents a Boolean function f_g, which maps k input bits to one output bit, i.e., $g(w_1, w_2, \ldots, w_k) : \{0, 1\}^k \rightarrow \{0, 1\}$. The most important unary gate is the NOT gate ($\neg$), while typical binary gates are AND ($\wedge$), OR ($\vee$), and XOR ($\oplus$). Moreover, we distinguish between linear[1] (e.g., XOR) and non-linear (e.g., AND, OR) gates, as they have different evaluation costs in secure computation [49]. In some works on MPC, non-linear gates are also referred to as non-XOR gates.

Notation For a function f, we refer to its circuit representation as C_f. We use s to denote the total number of gates in a circuit, also referred to as size, i.e., $s = size(C_f)$. Moreover, we use s^{nX} to count the number of non-linear (non-XOR) gates

[1]The Boolean function represented by linear gates can be expressed by a linear combination of its inputs over $\mathbb{Z}_2$, e.g., $XOR(X, Y) = X + Y \mod 2$.

N. Büscher, S. Katzenbeisser, *Compilation for Secure Multi-party Computation*, SpringerBriefs in Computer Science, https://doi.org/10.1007/978-3-319-67522-0_2

per circuit. We denote the depth of a circuit as d and use d^{nX} to denote its non-linear depth of a circuit, which is the maximum depth of all output gates, defined recursively for a gate g as:

$$d^{nX}(g) = \begin{cases} 0 & \text{if } g \text{ is input gate,} \\ max(d^{nX}(w_1), d^{nX}(w_2)) & \text{if } g \text{ with inputs } w_1, w_2 \text{ is linear,} \\ max(d^{nX}(w_1), d^{nX}(w_2)) + 1 & \text{if } g \text{ with inputs } w_1, w_2 \text{ is non-linear.} \end{cases}$$

Furthermore, we denote bit strings in lower-case letters, e.g., x. We refer to individual bits, which can be part of a bit string, in capital letters X and denote their negation with $\overline{X}$. We refer to a single bit at position i within a bit string as X_i. The Least-Significant Bit (LSB) is denoted with X_0. When writing Boolean equations, we denote AND gates with $\cdot$, OR gates with $+$ and XOR gates with $\oplus$. Moreover, when useful, we abbreviate the AND gate $A \cdot B$ with AB.

Integer Representation Integers are represented in binary form, as typical in computer science and logic synthesis. Hence, the decimal value of a bit string x is $\sum_{i=0}^{n-1} 2^i \cdot X_i$. Throughout this book, we use the two's complement representation to represent signed binary numbers. This common representation has the advantage that arithmetic operations for unsigned binary numbers such as addition, subtraction, and multiplication can be reused for signed numbers. In the two's complement, negative numbers are represented by flipping all bits and adding one: $-x = \overline{x} + 1$. In the following sections, we assume a two's complement representation when referring to negative numbers.

Fixed and Floating Point Representation To represent real numbers, we use two representations. The fixed point representation has a fixed position of the radix point (the decimal point). Hence, a binary bit string is split into an integer part and a fractional part. The decimal value of a bit string in fixed point representation is computed as $\sum_{i=0}^{n-1} 2^{i-r} \cdot X_i$, where r is the position (counted from the LSB) of the radix point. The IEEE-754 floating point representation is the standard representation of floating point numbers. It divides a bit string in three components, namely sign s, significant m and exponent e. Its real value is determined by computing $(-1)^s \cdot m \cdot 2^e$. In the IEEE-754 standard, the bit-width of each component, as well as their range is determined. Moreover, various error handling behavior is specified, e.g., overflow or division by zero [36].

2.2 Secure Computation

Secure multi-party computation (MPC) has been proposed in the 1980s as a more theoretical construct [75]; it only became practical in the last decade. As we focus on compilation in this book and assume that the reader has some familiarity with MPC, we do not give an in depth description of MPC protocols. However, we give a high-level description of the two first and most prominent MPC protocols over Boolean circuits, namely *Yao's garbled circuits* and the *GMW* protocol, whose evaluation costs are quite prototypic for most subsequently proposed constant and multi-round MPC protocols.

We explain all protocols in the *semi-honest* (passive) adversary model. Protocols secure against semi-honest adversaries ensure correctness and guarantee that participating parties do not learn more about the other party's inputs than they could already derive from the observed output of the joint computation. The semi-honest model is opposed the *malicious model*, where the adversary is allowed to actively violate the protocol. Typically, MPC protocols in the semi-honest model are the building blocks for protocols secure against malicious adversaries and are also used for many privacy-preserving applications on their own.

We begin with an introduction into oblivious transfer, which is required for both Yao's garbled circuits and GMW.

2.2.1 Oblivious Transfer

An oblivious transfer protocol (OT) is a protocol in which a sender transfers one of multiple messages to a receiver, but it remains oblivious which piece has been transferred. At the same time, the receiver can select the message that he wants to retrieve. In this chapter, we only use 1-out-of-2 OTs, where the sender inputs two l-bit strings m_0, m_1 and the receiver inputs a bit $c \in \{0, 1\}$. At the end of the protocol, the receiver obliviously receives m_c such that neither the sender learns the choice c, nor the receiver learns anything about the other message m_{1-c}. OT protocols require asymmetric cryptography and where assumed to be very costly in the past. However, in 2003 Ishai et al. [46] presented the idea of *OT Extension*, which significantly reduces the computational costs of OTs for most interesting applications of MPC. Recent implementations of OT Extension are capable of performing multiple millions of OTs per second on a single core [1, 47].

2.2.2 Yao's Garbled Circuits Protocol

Yao's garbled circuits protocol, proposed in the 1980s [76], is the most well-known secure two-party computation protocol, secure in the semi-honest model. The protocol is run between two parties P_A, P_B and operates on functionality descriptions in form of Boolean circuits over binary Boolean gates. To securely evaluate a functionality $f(x, y)$ over their private inputs x and y, both parties agree on a circuit $C_f(x, y)$, which can be seen as the machine code for the protocol.

During protocol execution, one party becomes the circuit generator (the garbling party), the other the circuit evaluator. The generator initializes the protocol by assigning each wire w_i in the circuit two random labels w_i^0 and w_i^1 of length κ

(the security parameter), representing the respective Boolean values 0 and 1. For each gate the generator computes a *garbled truth table*. Each table consists of four encrypted entries of the output wire labels w_o^γ. These are encrypted according to the gate's Boolean functionality $g(\alpha, \beta)$ using the input wire labels w_l^α and w_r^β as keys. Thus, an entry in the table is encrypted as

$$E_{w_l^\alpha}(E_{w_r^\beta}(w_o^{g(\alpha,\beta)})).$$

After their creation, the garbled tables are randomly permuted and sent to the evaluator, who, so far, is unable to decrypt a single row of any garbled table due to the random choice of wire labels. To initiate the circuit evaluation, the generator sends its input bits x in form of his input wire labels to the evaluator. Moreover, the wire labels corresponding to the evaluator's input y are transferred via an OT protocol, with the generator being the OT sender, who inputs the two wire labels, and evaluator being the OT receiver, who inputs it input bits of the computation. After the OT step, the evaluator is in possession of the garbled circuit and one input label per input wire. With this information the evaluator is able to iteratively decrypt the circuit from input wires to output wires. Once all gates are evaluated, all output wire labels are known to the evaluator. In the last step of the protocol, the generator sends an output description table to the evaluator, containing a mapping between output label and actual bit value. The decrypted output is then shared with the generator.

Implementations and Optimizations Yao's original protocol has seen multiple optimizations in the recent past. Most important are *point-and-permute* [4, 76], which allows an efficient permutation of the garbled table such that only one entry in the table needs to be decrypted by the evaluator, *garbled-row-reduction* [64], which reduces the number of ciphertexts that are needed to be transferred per gate, *free-XOR* [49], which allows to evaluate linear gates (XOR/XNOR) essentially for 'free' without any encryption or communication costs, and finally the communication optimal *half-gate* scheme [78], which requires only two ciphertexts per non-linear gate while being compatible with free-XOR. Important for practical implementations is the idea of *pipe-lining* [43], which is necessary for the evaluation of larger circuits and a faster online execution of Yao's protocol, as well as the idea of *fixed-key garbling* [5], which allows to achieve garbling speeds of more than 10M gates per second on a single CPU core using the AES instructions of modern CPUs.

Cost Model Considering all optimizations mentioned above, then for a given Boolean function $f(x, y)$ and its circuit representation $C_f(x, y)$, the evaluation costs of a circuit in Yao's protocol are dominated by the *number of the input bits*, and by the *number of non-linear gates in the circuit*. The evaluation costs of linear gates are for most applications negligible, as communication is the current practical bottleneck of Yao's protocol. To garble a non-linear gate, two entries from the garbled table of length κ have to be transferred. Assuming a standard key length of $\kappa = 128$ bit, then a garbling throughput of ten million non-linear gates per second produces $\approx$2.8 Gbit of data per second. Considering that the communication requirement of two ciphertexts per gate is a proven lower bound for known garbling schemes [78], minimizing the number of non-linear gates in a circuit is an important optimization goal when creating applications (circuits) for Yao's protocol. We remark that this optimization goal holds for almost all MPC protocols over Boolean circuits, as these have a similar mechanism to free-XOR, which allow the evaluation of linear gates without communication.

2.2.3 Goldreich-Micali-Wigderson (GMW) Protocol

The *Goldreich-Micali-Wigderson (GMW)* protocol [37] also originates from the 1980s and allows two parties to securely compute a functionality described in form of a Boolean circuit. In contrast to Yao's protocol, which uses the idea of garbled tables, GMW is built on top of Boolean secret sharing. Each input and intermediate wire value is shared among the two parties using an XOR based sharing scheme over single bits, i.e., for every value $V \in \{0, 1\}$ each party $P \in \{0, 1\}$ holds a share V^P, indistinguishable from random, with $V = V^0 \oplus V^1$. Values can be revealed at the end of the computation by exchanging and XORing the shares.

Given shared values, XOR gates can be evaluated locally without any communication between the parties by XORing the respective wire shares, e.g., both parties compute $W^P = U^P \oplus V^P$ over their shares of U and V to compute $W = U \oplus V$. AND gates require the parties to run an interactive protocol, which can be realized with Boolean multiplication triples [3]. A multiplication triple consists of three bits A, B, C, where $C = A \cdot B$ holds, that are shared between the parties. Given such a triple, the two parties can compute an AND gate $Z = X \cdot Y$, by opening (reconstructing) two bits $E = A \oplus U$ and $F = B \oplus V$ and computing their shares of Z as

$$Z^P = P \cdot E \cdot F \oplus F \cdot A^P \oplus E \cdot B^P \oplus C^P .$$

Implementations and Optimizations A first implementation was given by Choi et al. [20] that was subsequently be improved by Schneider et al. [71]. Furthermore, Asharov et al. [1] showed that the triples can be precomputed using only two random OTs with a total communication of $2 \cdot \kappa$ bits, where κ denotes the key length in bits. A protocol secure against malicious adversaries, named TinyOT, which is based on GMW, was proposed by Nielsen et al. [66].

Cost Model The multiplication triples, which are the most expensive part of the computation, can be precomputed (known as the offline phase), as they are independent of the actual functionality to be computed. This yields a communication complexity in the online phase of four bits (two per party) for every AND gate. All AND gates on the same layer in the circuit can be computed in parallel, hence, in the same communication round. We observe, that in practice, the computation effort of AND and XOR gates is negligible when compared to the communication costs. In summary, the performance of the GMW protocol for a given circuit depends on two aspects, namely the total number of non-linear gates s^{nX} as well as the non-linear depth d^{nX} of the circuit (number of layers of AND gates), which is prototypic for multi-round MPC protocols over Boolean circuits.

2.3 Benchmarking Applications for MPC Compilers

MPC has many applications, e.g., privacy-preserving biometric authentication [31], private set intersection [33], or secure auctions [13]. For research purposes, some (parts of) these applications have become popular for benchmarking purposes. The most prominent example is the AES block cipher, which is the de facto standard benchmark for MPC protocols. In compiler research on MPC, multiple different functionalities are commonly compiled into circuits, which are then compared regarding their properties, i.e., size, depth and fraction of non-linear gates.

In this section, we give an overview of these commonly benchmarked functionalities, ranging from very small snippets, e.g., integer addition, to larger functionalities, e.g., IEEE-754 compliant floating point operations or biometric authentication. In later chapters, we use these functionalities to evaluate the techniques presented in this book. Here, we discuss the functionalities together with their complexity and motivate why these functionalities are relevant in the context of MPC and interesting for the development of compilers.

Common benchmark functionalities are:

- *Arithmetic building blocks.* Due to their (heavy) use in almost every computational problem, arithmetic building blocks are of high importance in high-level circuit synthesis and are often benchmarked individually. Most common building blocks are addition, subtraction, multiplication, and division. When studying these building blocks, one should distinguish the results for different input and output bit widths. Namely, arithmetic operations can be differentiated in overflow-free operations, e.g., allocating $2n$ bits for the result of a $n \times n$ bit multiplication, and operations with overflow, e.g., when only allocating n bits for the same multiplication. In later chapters we explicitly state which input and output bit-widths are studied. To evaluate the scalability of compilers, multiple sequentially (in-)dependent arithmetic operations can be compiled as a single functionality.
- *Matrix multiplication.* Algebraic operations, such as matrix or vector multiplications, are building blocks for many privacy-preserving applications, e.g., for feature extractors in biometric matching or the privacy friendly evaluation of neural networks, and have therefore repeatedly been used to benchmark compilers for MPC [28, 42, 52]. From an optimizing compilers perspective, matrix multiplication is a comparably simple task, as it only involves the instantiation of arithmetic building blocks. However, it is very well suited to show the

scalability of compilers or the capability to fuse multiple arithmetic operations into a single statements, as required for depth minimization (see Chap. 6). Matrix multiplication can be parametrized according to the matrix dimensions, the used number representation (integer, fixed or floating point) and its bit-width.

- *Modular exponentiation.* Modular exponentiation, i.e., $x^y \bmod p$, is relevant in secure computation, as it is a building block for various cryptographic schemes. For example, blind signatures can be realized with RSA using modular exponentiation. Blind signatures allow a signing process, where neither the message is revealed to the signing party, nor the signing key is revealed to the party holding the message. Therefore, modular exponentiation has been used multiple times to study the performance of MPC [5, 29, 51].
- *Distances.* Various distances need to be computed in many privacy preserving protocols and are therefore relevant for MPC, e.g., in biometric matching applications or location privacy applications. The *Hamming* distance is a measure between two bit strings. Computed is the number of pairwise differences in every bit position. Due to its application in biometrics, the Hamming distance has often been used for benchmarking MPC compilers, e.g., [42, 51, 61]. The *Manhattan* distance $dist_{MH} = |x_1 - x_2| + |y_1 - y_2|$ between two points $a = (x_1, y_1)$ and $b = (x_2, y_2)$ is the distance along a two dimensional space, i.e., along the Manhattan grid, when only allowing horizontal or vertical moves. The *Euclidean* distance is defined as $dist_{ED} = \sqrt{(x_1 - x_2)^2 + (y_1 - y_2)^2}$. Due to the complexity of the square root function, it is common in MPC to benchmark the squared Euclidean distance separately, as its computation is usually sufficient when comparing multiple distances. All distances can be parameterized according the used input bit-widths.
- *Biometric matching.* In biometric matching a party matches one biometric sample (a vector of features) against the other's party database of biometric templates. Example scenarios are face-recognition or fingerprint-matching [31]. One of the main concepts is the computation of a distance (see above) between the sample and all database entries. Once all distances are computed, the minimal distance determines the best match. The biometric matching application is very interesting for benchmarking, as it involves many parallel arithmetic operations, as well as a large number of sequential comparisons. The biometric matching application can be parameterized by the database size, the sample dimension (number of features per sample), the bit-width of each feature, and the used distance.
- *Location-aware scheduling.* Privacy-preserving availability scheduling is another application that can be realized with MPC [10]. The functionality matches the availability of two parties over a number of time slots, without revealing the individual schedule to the other party. Location-aware scheduling also considers the location and maximum travel distance of the two parties for a given time slot. Therefore, the functionality outputs a matching time slot where both parties are available and in close proximity to each other (if existent). The functionality can be parameterized by the number of time slots used per party, the representation of locations, and computation of distances, e.g., Manhattan or Euclidean distance.

- *Median computation.* The secure computation of the median is required in privacy preserving statistics. It is an interesting task for compilation, as it can be implemented using a sorting algorithm, because a sorted array allows a direct access to the median element. The compiled circuit depends on the used sorting algorithm. In this book, we evaluate Bubble and Merge sort. Moreover, the task can be parameterized according number of elements and their bit-width.
- *Fixed and floating point arithmetics.* Fixed and floating point number representations allow the computation on real numbers using integer arithmetics. Therefore, these representations are necessary for all applications where numerical precision is required, e.g., in privacy preserving statistics. The floating point representation is the more versatile representation as it allows to represent a larger range of values, whereas fixed point arithmetics can be realized with significant less costs and are therefore of interest in MPC. Floating point operations are also very suited to evaluate the gate level optimization methods of compilers since they require many bit operations. When implementing floating point arithmetics, we follow the IEEE-754 standard.

Implementation Differences We remark that for a fair comparison between multiple compilers that compile the same functionality, it has to be made sure that the same algorithm as well as the same abstraction level of the source code is used. To illustrate this thought we give three example implementations for computing the *Hamming* distance, which produce circuits of largely varying sizes (cf., Sect. 4.5). The distance can be computed by XOR-ing the input bit strings and then counting the number of bits. An exemplary implementation is given in Listing 2.1 that computes the distance between two bit-strings of length 160 bit, which are split over five unsigned integers. In Line 8 the number of ones in a string of 32 bits is computed. This task is also known as population count. In the following paragraphs we describe three different implementations.

```
1  #define N 5
2
3  void hamming() {
4    unsigned INPUT_A_x[N];
5    unsigned INPUT_B_y[N];
6    unsigned res = 0;
7    for(int i = 0; i < N; i++) {
8      res += count_naive32(INPUT_A_x[i]^INPUT_B_y[i]);
9    }
10   unsigned OUTPUT_res = res;
11 }
```

Listing 2.1 Hamming distance computation between two bit strings

The first implementation is given in Listing 2.2. In this naïve approach, each bit is extracted using bit shifts and the logical AND operator before being aggregated.

```
unsigned char count_naive32(unsigned y) {
  unsigned char m = 0;
  for(unsigned i = 0; i < 32; i++) {
    m += (y & (1 << i)) >> i;
  }
  return m;
}
```

Listing 2.2 Counting bits using a naïve bit-by-bit comparison approach

A variant of this implementation is given in Listing 2.3. Here, the bit string of length 32 is first split into chunks of 8 bits (`unsigned char`). The ones set in each chunk are then counted as described above.

```
unsigned char count_naive8(unsigned char c) {
  unsigned char m = 0;
  for(int i = 0; i < 8; i++) {
    m += (c & (1 << i)) >> i;
  }
  return m;
}

unsigned char count_tree32(unsigned y) {
  unsigned char m0 = y & 0xFF;
  unsigned char m1 = (y & 0xFF00) >> 8;
  unsigned char m2 = (y & 0xFF0000) >> 16;
  unsigned char m3 = (y & 0xFF000000) >> 24;
  return count_naive8(m0) + count_naive8(m1) + \
    count_naive8(m2) + count_naive8(m3);
}
```

Listing 2.3 Counting bits over unsigned chars in a tree based manner

Finally, in Listing 2.4 a variant optimized for a CPU with 32 bit registers and slow multiplication is given. This implementation uses only 14 CPU instructions.

```
unsigned count_reg32(unsigned y) {
  unsigned x = y - ((y >> 1) & 0x55555555);
  x = (x & 0x33333333) + ((x >> 2) & 0x33333333);
  x = (x + (x >> 4)) & 0x0f0f0f0f;
  x += x >>  8;
  x += x >> 16;
  return x;
}
```

Listing 2.4 Counting bits, optimized for a 32 bit register machine

Fundamental implementation differences will inevitable lead to different compilation results. Therefore, all benchmarked functionalities in this book are implemented using the same algorithms, data structures and data types when using them for the comparison with other compilers using possibly different input languages.

Chapter 3
Compiling ANSI-C Code into Boolean Circuits

3.1 Motivation and Overview

When visualizing an MPC protocol as a hardware architecture that can execute programs (functionalities), then these programs have to be described in form of Boolean circuits (e.g., for Yao's garbled circuits [76] or the GMW protocol [37]) or Arithmetic circuits (e.g., for the Sharemind [12] or the SPDZ protocol [26]). In this book, we focus on the case of Boolean circuit based MPC protocols. We show how compilers can be constructed that take a source code, written in a high level language, as input and convert the description into an equivalent Boolean circuit representation suitable for MPC. In this way, compilers for MPC help to abstract from cryptographic protocols as well as classic hardware design.

In this chapter, we show which steps are necessary to automatically convert a given input source code into a Boolean circuit. We do this by explaining the functionality of our compiler *CBMC-GC* that allows developers without a professional background in computer security and hardware design to develop efficient applications for MPC.

Technically, CBMC-GC is based on the software architecture of the model checker CBMC by Clarke et al. [21], which was designed to verify ANSI C source code. Using a well tested bit-precise model checker allows for a very reliable compilation chain from C to circuits for MPC. CBMC transforms an input C program f, including assertions that encode properties to be verified, into a Boolean formula B_f, which is then analyzed by a SAT solver. The formula B_f is constructed in such a way that the Boolean variables correspond to the memory bits manipulated by the program and to the assertions in the program. CBMC is an example of a *bit-precise* model checker, i.e., the formula B_f encodes the real life memory footprint of the analyzed program on a specific hardware platform under ANSI C semantics. The construction of the formula B_f ensures that satisfying assignments found by the SAT solver are program traces that violate assertions in the program.

N. Büscher, S. Katzenbeisser, *Compilation for Secure Multi-party Computation*, SpringerBriefs in Computer Science, https://doi.org/10.1007/978-3-319-67522-0_3

Thus, CBMC is essentially a compiler that translates C source code into Boolean formulas. The code must meet some requirements, detailed in Sect. 3.3.1, so that this transformation is possible in an efficient manner. In particular, the program must terminate in a finite number of steps; CBMC expects a number k as input which bounds the size of program traces (and CBMC also determines if this bound is sufficient). The compiler CBMC-GC inherits these constraints from CBMC. However, for MPC this property is actually a mandatory requirement rather than a limitation. This is because combinatorial circuits have a fixed size and thus deterministic evaluation time in any MPC protocol. This is a logical consequence from the requirement that the runtime of a MPC protocol should not leak any information about the inputs of either party.

ANSI C As Input Programming Language In contrast to many other compilers for MPC (see Sect. 1.2), CBMC-GC does not use a new domain-specific programming language or even a hardware description language. Instead CBMC-GC uses standardized ANSI C as input language. This has the significant advantage that existing code can be reused, and that application developers do not need to learn a new language to explore MPC. For example, in the later chapters we study the use of floating point numbers in MPC, whose support can be added to CBMC-GC by compiling one of the many existing software floating point implementations. The caveats of compiling a common programming language, where for example the bit-widths of the data types do not match the bit-widths on the circuit level, are addressed in detail the next chapter, where we introduce various optimization methods.

At first sight, choosing ANSI C may look surprising: While C has a standardized ANSI semantics, C is not the latest fashion in programming languages, it does not achieve the platform independence of Java, and it is also lacking advanced programming features such as object orientation. While these objections are valid, they are outweighed by technical and practical arguments in favor of C:

- The C language is quite close to the underlying hardware, and a competent programmer has good control of the actual computation on the processor. In particular, the programmer can control the memory and CPU use of his program better than in other high level language. This proximity to the hardware platform has made C the language of choice for areas like embedded systems, device drivers and operating systems. For similar reasons, hardware vendors are customarily prototyping hardware components in C.
- As we demonstrate in the next chapter, this advantage of C translates to MPC. The sizes of the circuits that we obtain from compiling C programs are surprisingly small. Since the practicality of MPC is heavily dependent on the circuit size, we conclude that with current technology, C is a high level language very suitable for our task.
- While the semantics of C may be less than pretty to the eye of programming language theorists, the conceptual simplicity of C makes automated safety analysis of C programs (e.g., by software model checking and static analysis) much simpler than for languages such as Java and C++.

- The availability of software verification tools for C has the additional advantage that we can address correctness of the program in a systematic way. We note that the limited size of programs realizable in MPC makes them amenable for code analysis tools.

Chapter Outline Next, we describe the basics of model checking in Sect. 3.2. Then we introduce our compilation chain in Sect. 3.3. Finally, in Sect. 3.4 we describe the computational costs of ANSI C operations in an MPC circuit.

3.2 Background: Bounded Model Checking

Safety analysis of software and hardware has become very practical in the last 15 years. One of the analysis techniques is *model checking*, which is used to verify correctness properties of a system by checking whether a model of the system meets a given specification. For example, such a property can be a safety requirement, e.g., the absence of deadlocks, in embedded devices deployed in critical infrastructures. Model checking functions by automatic exhaustive search through the state space of the model, which has to be finite.

The basic idea of *bounded model checking (BMC)* [8] is to search for a counterexample in executions whose length is bounded by some integer k. The BMC problem can be efficiently reduced to a propositional satisfiability problem (SAT), and can therefore be solved by SAT methods. One flavor of bounded model checking is bit-precise reasoning. In bit-precise reasoning, the program semantics is precisely modeled in a suitable logical formalism, most importantly in Boolean logic. While other flavors, e.g., overapproximating model checkers, often model integer variables by unbounded (mathematical) integer values, a bit-precise model checker will typically model them as a bit vector with real-life overflow behavior. The most well-known bit-precise model checkers are DiVer[35] and CBMC [21], where CBMC is used in this work.

For a given bound k, the model checker CBMC transforms a program written in ANSI C into a Boolean constraint whose solutions are program traces of size at most k which violate one or more assertions in the program. These solutions are then determined by a Boolean SAT solver. Importantly, this transformation identifies the bits manipulated during program execution with Boolean variables in the formula. Notwithstanding the high theoretical complexity of SAT solving, modern SAT solvers [9] can often solve verification constraints with millions of clauses. Note that in common terminology, CBMC can also be classified as an instance of SAT-based model checking and bounded model checking [8].

The important property of CBMC for our purposes is its capability to generate bit-precise Boolean descriptions of the program execution from the source code. This capability forms the basis for the CBMC-GC compiler which we describe in the next section.

3.3 CBMC-GC's Compilation Chain

CBMC-GC's compilation pipeline, which is in most parts using CBMC's original compilation pipeline, is illustrated in Fig. 3.1. A given input source code passes through multiple steps, before being converted into a circuit:

1. *Parsing and type checking.* First, CBMC-GC parses the given source code into a parse tree and checks syntactical correctness of the source code.
2. *GOTO conversion.* Then, the code is translated into a GOTO program, which is the intermediate representation of code in CBMC-GC.
3. *Loop unrolling.* Next, the program is made loop free by unrolling all loops and recursions.
4. *Single-static assignment (SSA).* The loop-free program is rewritten in SSA form, where every variable is only assigned once.
5. *Expression simplification.* Using symbolic execution techniques, constants are propagated and expressions are simplified in the SSA form.
6. *Circuit instantiation.* Finally, given the simplified SSA form, a Boolean circuit is instantiated.

In this section, we first describe the differences between standard C code and code for CBMC-GC. Then, we give an overview of all the compilation steps mentioned before and explain them in detail. Moreover, we give code examples and point out the differences between compiling a program into a SAT formula (CBMC) and compiling into a logical circuit for MPC (CBMC-GC).

3.3.1 Input Language and Circuit Mapping

When programming for CPU/RAM architectures, inputs and outputs of a program are commonly realized with standard libraries that themselves invoke system calls of the operating system. Contrasting, in MPC the only input and output interface available are the *input/output (I/O)* wires of the circuit. To realize the I/O mapping between C code and circuits, we use a special naming convention of I/O variables. Furthermore, the input variables have to be left uninitialized in the source code and

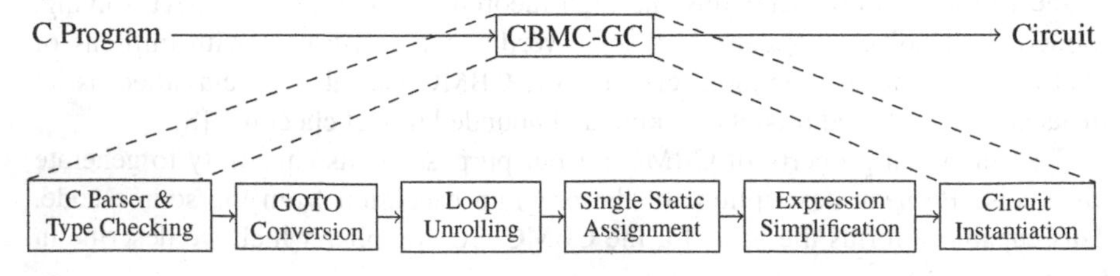

Fig. 3.1 Compilation chain. CBMC-GC's compilation pipeline without optimization

are only assigned a value during the evaluation of the circuit in an MPC framework. Hence, instead of adding additional standard libraries, CBMC-GC requires the developer to name input and output variables accordingly.

To illustrate this naming convention, we give an example source code of the millionaires' problem in Listing 3.1. The function shown is a standard C function, where only the input and output variables are specifically annotated as designated input of party P_A or P_B (Lines 2 and 3) or as output (Line 4). Hence, variables that are inputs of party P_A or P_B have to be named with a preceding `INPUT_A` or `INPUT_B`. Similar, output variable names have to start with `OUTPUT`. Aside from this naming convention, arbitrary C computations are allowed to produce the desired result, in this case a simple comparison (Line 6).

```
1 void millionaires() {
2   int INPUT_A_income;          // Input Party A
3   int INPUT_B_income;          // Input Party B
4   int OUTPUT_result = 0;       // Output
5
6   if (INPUT_A_income > INPUT_B_income) {
7     OUTPUT_result = 1;
8   }
9 }
```

Listing 3.1 Millionaires' problem. CBMC-GC code example for Yao's Millionaires' problem

For simplicity, CBMC-GC only distinguishes between two parties and uses a shared output, which is the simplest case of secure two-party computation. However, this is not preventing the compilation of source code for more than two parties or code with outputs that are designated for specific parties only. This is because, during compilation CBMC-GC only distinguishes input and output variables, but not the association with any party. Hence, to compile code for more parties, an application developer can use its own naming scheme that extends the one introduced by CBMC-GC. For example, to use different outputs for three parties the following convention could be used: `OUTPUT_A_res`, `OUTPUT_B_res` and `OUTPUT_C_res`.

CBMC-GC outputs a mapping between every I/O variable and their associated wires in the circuit; this information can be used in any MPC protocol implementation to correctly map wires back to the designated parties.

3.3.2 C Parser and Type Checking

CBMC-GC parses the given source code using standard compilation techniques, e.g., using an off-the-shelf C preprocessor (e.g., `gcc -E`) that implements the macro language of C. The preprocessed code is then parsed using a common lexer and parser setup, namely GNU `flex` and `bison` to parse the code into an abstract syntax tree (AST) representation. During parsing, a symbol table is created, which tracks all occurring symbols and their bit-level type information. Type checking is already performed during parsing using the symbol table. If any inconsistencies are detected, CBMC-GC will abort the compilation. The whole parsing process resembles a typical compiler *frontend*, as it is also required when compiling for CPU/RAM architectures. More background on this part of the compile chain can be found in [63].

3.3.3 GOTO Conversion

In the second phase of the compilation chain, the parsed AST is translated into a *GOTO* program, which is the intermediate representation of code in CBMC-GC. In this representation all operations responsible for non-sequential control flow, such as `for`, `while`, `if`, or `switch` statements are replaced by equivalent *guarded* goto statements. These statements can be seen as if-then-else statements with conditional jumps, similarly to conditional branches in assembly language. Using a GOTO representation allows for a uniform treatment of all non-sequential control flow, i.e., no distinction between recursion and loops has to be made, which is beneficial for the next compilation step to unroll all loops. We also remark that in this phase CBMC-GC is capable of handling complex language features, such as function pointers, which are resolved using static analysis to identify and branch all candidate functions.

Goto Conversion Example To illustrate the conversion of source code into a GOTO program, we study the example code given in Listing 3.2. In this code, two types of loops are used, which will have the same representation in the GOTO program.

```
1  int main() {
2    int i, x;
3    i = 0;
4    while(x < 100) {
5      x *= 2;
6    }
7    for(i = 1; i < 5; i++) {
8      x = x * (x - 1);
9    }
10   return x;
11 }
```

Listing 3.2 Code example. Code snippet with one `for` and one `while` loop

The resulting control flow graph (CFG) after the conversion into a GOTO program is shown in Fig. 3.2. We observe that both loops have been replaced by a conditional branch ending in a GOTO statement. Recursive functions are translated in the same manner.

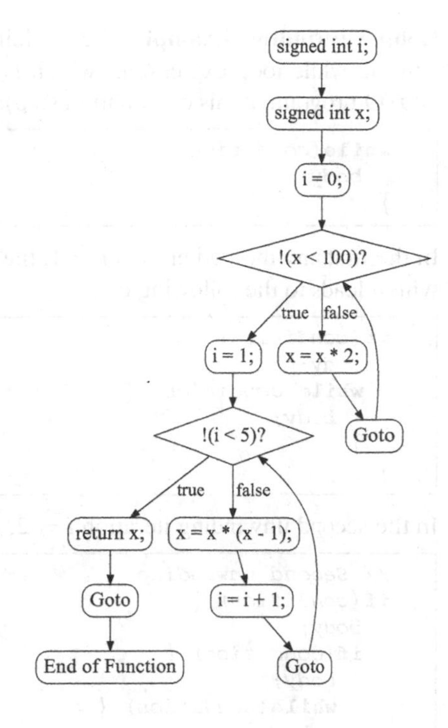

Fig. 3.2 Exemplary GOTO program. Shown is the CFG using the GOTO representation as generated by CBMC for the example code in Listing 3.2

3.3.4 Loop Unrolling

Boolean formulas and circuits are purely combinatorial, i.e., they only consist of Boolean operators (gates). Cyclic structures, which are still present in the GOTO program, have to be removed during compilation, because these cannot be represented in combinatorial form. To make the program acyclic, CBMC-GC performs symbolic execution to (greedily) *unwind* all loops and recursion in the program. Unwinding is done until either a termination of the cycle is detected during symbolic execution, e.g., the range of a `for` loop has been exceeded, or a user specified bound k has been reached. This bound can be set locally for every function, or globally for the complete program.

Technically, unwinding works by replacing all loops (cyclic GOTOs) by a sequence of k nested `if` statements. In CBMC the sequence is followed by a special assertion (called *unwinding assertion*) which can be used to detect insufficient k. For CBMC-GC, the assertion is omitted because it cannot be checked during circuit evaluation and therefore, the programmer has to ensure a correct upper bound for all loops. Automatically deriving an upper bound is impossible in many cases, as this relates to the undecidable halting problem. Thus, instead of restricting the input language, which would be the alternative solution, CBMC-GC's approach allows the programmer to use any cyclic code structures, as long as an upper bound for every cycle can be provided.

Loop Unwinding Example We explain the process of loop unwinding using a generic while loop expression (which is easier to read, yet actually replaced by a GOTO program at this compilation step):

```
while(condition) {
  body;
}
```

In the first unwinding iteration $i = 1$, the first iteration of the loop body is unrolled, which leads to the following code:

```
if(condition) {
  body;
  while(condition) {
    body;
  }
}
```

In the second unwinding iteration $i = 2$, the second loop iteration is unrolled:

```
// Second unwinding
if(condition) {
  body;
  if(condition) {
    body;
    while(condition) {
      body;
    }
  }
}
```

In the last iteration $i = k$, the loop is completely unrolled:

```
if(condition) {
        body;
        if(condition) {
                body;
                [...]
                if(condition) {
                        body;
                }
                [...]
        }
}
```

To provide a concrete example, the loop given below:

```
int x[3];
int i = 0;
while(i < 3){
  x[i] = x[i] * x[i];
  i = i + 1;
}
```

is unrolled to the following code:

```
int x[3];
int i = 0;
if(i < 3) {
  x[i] = x[i] * x[i];
  i = i + 1;
  if(i < 3) {
    x[i] = x[i] * x[i];
    i = i + 1;
    if(i < 3) {
      x[i] = x[i] * x[i];
      i = i + 1;
    }
  }
}
```

We remark that recursive functions can be handled in the same way as loops are unrolled by expanding the function body k times, as recursion and loops use the same representation in a GOTO program. We also observe that for loops with at most k steps, unwinding preserves the semantics of the program.

3.3.5 Conversion into Single Static Assignment Form

Once the program is acyclic, CBMC-GC turns it into SSA form. This means that each variable x in the program is replaced by fresh variables $x_1, x_2, \ldots$, where each of them is assigned a value only once. For instance, the code sequence

```
x = x + 1;
x = x * 2;
```

is replaced by

```
x2 = x1 + 1;
x3 = x2 * 2;
```

Conditionals occurring in a program are translated into *guarded assignments*. The core idea is that instead of branching at every conditional, both branches are evaluated on distinct program variables. After both branches have been evaluated, the effect on the variables, which are possibly modified by either of the two branches, is applied depending on the branch condition, also referred to as the guard. For example, the body of the following function, which computes the absolute of a value,

```
int abs(int x) {
  if (x < 0)
    x = x - 1;
  return x;
}
```

is transformed into guarded SSA form as follows:

```
int abs(int x1) {
  _Bool guard1 = (x1 < 0);
  int x2 = -x1;
  int x3 = guard1 ? x2 : x1
  return x3;
}
```

The SSA representation has the important advantage that we can now view the (guarded) assignments of program variables as mathematical equations. Note that, the interpretation of an assignment as equation, e.g., `x=x+1;` is unsolvable. The indices of the variables in SSA form essentially correspond to different intermediate states in the computation.

3.3.6 Expression Simplification

Expression simplification, e.g., constant folding, and constant propagation allow to perform computation on constants and to remove unnecessary computations already during compile time. Hence, removed operations do not need to be translated into circuits. For example, the expressions

```
x = a + 0;
y = a * 0 + b * 1;
z = -3 + a + 6 / 2;
```

can be simplified to

```
x = a;
y = b;
z = a;
```

by reordering the expressions and evaluating constant parts or by template based matching, such as rewriting `0 * X` by `0`. In Sect. 3.4, we will show that array accesses in MPC are comparably inefficient. Therefore, expression simplification is especially effective and important for the computation of array indices. Expression simplification can also be performed before loop unrolling, yet, in most cases more simplifications are possible after unrolling, as some variables will become constant, e.g., copies of a loop index variable.

3.3.7 Circuit Instantiation

In the final step, CBMC-GC translates the simplified code given in SSA form, which can be seen as a sequence of arithmetic equations, into Boolean formulas (circuits). To this end, CBMC-GC first replaces all variables by bit vectors. For instance, an

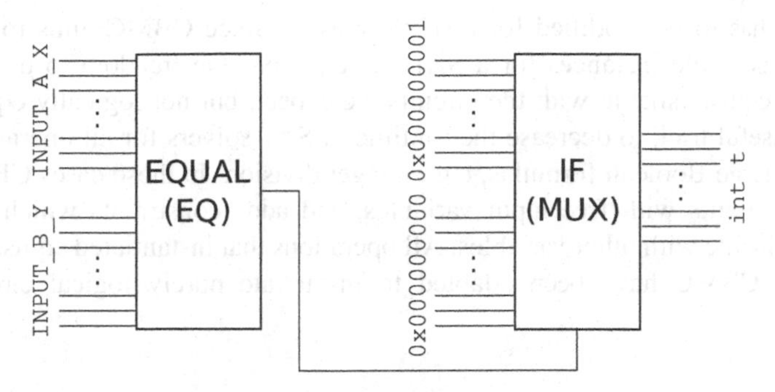

Fig. 3.3 Circuit (*left*: input, *right*: output) consisting of two Boolean functions, as created by CBMC-GC from the guarded assignment `(INPUT_A_X==INPUT_B_Y)? t=0 : t=1;`

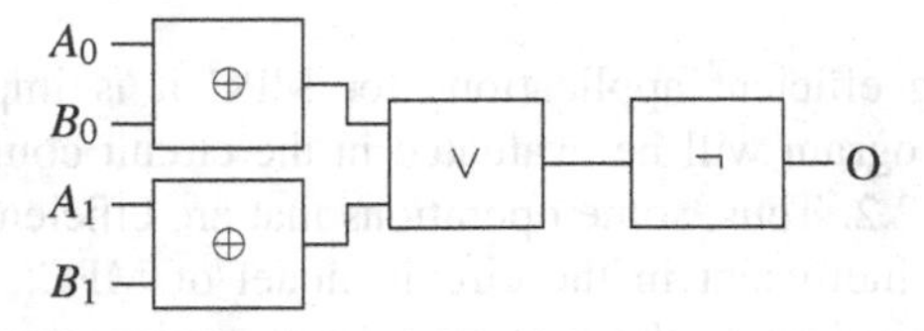

Fig. 3.4 The Boolean equivalence function for two input bit strings of length two, $a = A_0A_1$ and $b = B_0B_1$, implemented as a circuit consisting of two XOR, one OR and one NOT gate

integer variable will be represented as a bit vector consisting of 32 literals. For more complex variables such as arrays and pointers, CBMC-GC replaces all dereference operators by a function that maps all pointers to their dercferenced expression. More details can be found in [22].

Correspondingly, operations over variables (e.g., arithmetic computations) are naturally translated into Boolean functions over these bit vectors. Internally, CBMC-GC realizes these Boolean functions as circuits of Boolean gates whose construction principles are inspired by methods from hardware design. For example, the guarded assignment `(INPUT_A_X == INPUT_B_Y)? t = 0 : t = 1;` is translated into a circuit consisting of two Boolean functions, namely an equivalence check and a multiplexer. We refer to these elementary Boolean functions as *building blocks*. The resulting circuit in this example is illustrated in Fig. 3.3. An implementation of the equivalence building block is illustrated in Fig. 3.4 for a bit-width of two bits. In the next chapter, we will describe the construction of building blocks for all elementary operations. We also remark that for some operations, e.g., bit shifts or multiplication, building blocks for constant and variable inputs are distinguished during circuit instantiation. This is because operations with input known to be constant during compile time can be represented by a smaller circuit than those with variable input.

In the default setting, CBMC translates the resulting circuit into a Boolean formula in CNF form, ready for is use in a SAT solver. In CBMC-GC the translation into CNF form can be omitted. Moreover, at some places, the circuit generation of CBMC has to be modified for a subtle reason: Since CBMC aims to produce efficiently solvable instances for a SAT solver, it has the freedom to use circuits which are equisatisfiable with the circuits we expect, but not logically equivalent. This is a useful trick to decrease the runtime of SAT solvers for all operations that result in a large Boolean formulas, e.g., integer division. In these cases CBMC can introduce circuits with free input variables, and adds constraints which requires them to coincide with other variables. All operations that instantiated equisatisfiable circuits in CBMC have been adapted to instantiate purely logical circuits for CBMC-GC.

3.4 Complexity of Operations in MPC

When programming efficient applications for MPC it is important to consider that the resulting program will be evaluated in the circuit computation model, as described in Sect. 2.2.2. Thus, some operations that are efficient in the CPU/RAM model can be very inefficient in the circuit model of MPC. In this section, we give a high-level overview on the performance of various operations of ANSI C in the circuit model. A detailed discussion on how the different operations are implemented as a circuit and how these are optimized for MPC is given in the next chapter.

Table 3.1 shows a summary of the circuit complexities, i.e., the number of non-linear gates, which describe the computation costs in MPC, of various operations in ANSI C. Moreover, we illustrate the actual circuit sizes for a standard integer with a bit-width of $n = 32$ bit and give an estimate on the number of possible operations per second, when garbling each operation in Yao's protocol secure against semi-honest adversaries. The estimates are given using the fastest known garbling scheme [5], which garbles around 10,000,000 gates per second on single core of a commodity laptop CPU. We remark that already 4,000,000 non-linear gates per second are sufficient to saturate a network link with a capacity of 1Gbit [78].

Bit level operations in ANSI C, such as logical ANDs, OR, XOR are very efficient on the circuit level, as they directly translate into the according gate types with a total circuit size that is linear in the data type's bit-width (AND, OR) or even zero (XOR). Shifts (« or ») with variable offset are slightly more costly by a logarithmic factor. All logical operations do not require any gate at all, when being used with a constant.

Some arithmetic operations, like addition and subtraction, are likewise efficient with a circuit size linear in the bit-width. For example, it is possible to perform more than 300,000 additions per second on a single core in Yao's protocol. Multiplications and divisions are significantly more costly with a circuit size that is quadratic in the

Table 3.1 Circuit complexity of operations in ANSI C depending on the bit-width n in the number of non-linear gates and the size of an array m. Shown is also the size of each operation for a typical integer bit-width of $n = 32$ bits and an exemplary array size of $m = 1024$, as well as the number of possible operations in current state-of-the-art semi-honest Yao's garbled circuits on a single CPU core (10,000,000 non-linear gates per second [5])

	Circuit complexity	Complexity for 32 bit integer	Operations/second in Yao's protocol
Logical operators			
AND (&)	$O(n)$	32	312,500
OR (|)	$O(n)$	32	312,500
XOR (^)	$O(n)$	0	> 10M
Shift («/») variable	$O(n\log(n)))$	160	62,500
One of the above with constant	0	0	> 10M
Arithmetic operators			
Addition (+)	$O(n)$	31	322,580
Subtraction (−)	$O(n)$	32	312,500
Multiplication (*)	$O(n^2)$	993	10,070
Division (/)	$O(n^2)$	1085	9216
Modulo (%)	$O(n^2)$	1085	9216
Assignments			
var $\leftarrow$ var	0	0	> 10M
var $\leftarrow$ Array$_m$[const]	0	0	> 10M
Array$_m$[const] $\leftarrow$ var	0	0	> 10M
var $\leftarrow$ Array$_m$[var]	$O(mn)$	31,744	305
Array$_m$[var] $\leftarrow$ var	$O(mn)$	34,816	287

bit-width.[1] Hence, for a typical integer, multiplications are by a factor of 32 slower than additions. Nevertheless, In Yao's protocol more than 10,000 multiplications can be performed per second.

The assignment or copying of variables requires no gates in MPC, as assignments are represented by wires. The same holds for array access (read and write) with an index known at compile time. However, a main performance bottleneck in MPC is the access to an array with variable index. Each access (read or write) of an array element requires a circuit that has the size of the array itself. Assuming an array with m elements of bit-width n, a circuit for read or write access has size $O(mn)$. Consequently, for an exemplary, moderately sized array with '1024 integers, only 300 read or write access can be performed per second.

[1]For larger bit-widths a complexity of multiplication of $O(n^{1.6})$ can be achieved using Karatsuba multiplication [49].

In summary, integer arithmetics and logical operations are very fast, where multiplications and divisions are the slowest operations, yet still can be performed at a speed of 10,000 of operations per second. However, dynamic array access can become an obstacle for any practical application that requires arrays of noticeable size. Therefore, in the next chapter we also study techniques that aim at detecting constant array accesses to avoid a compilation of dynamic accesses whenever possible.

Chapter 4
Compiling Size-Optimized Circuits for Constant-Round MPC Protocols

4.1 Motivation and Overview

In the previous chapter we described a general compilation chain, based on the model checker CBMC to translate any (bounded) C program into a Boolean circuit serving as application description for MPC protocols. During protocol runtime, each gate in the circuit is evaluated in software using different cryptographic instructions depending on the gate types (see Sect. 2.2.2). Therefore, the performance of MPC applications depends both on the size of a circuit that represents the functionality to be computed and the gate types used therein.

We also observe that for almost all deployment scenarios of MPC protocols, the circuit is generated once, whereas the MPC protocol itself will be run multiple times. Especially when considering the high computational and communication costs of MPC protocols (compared to generic computation), it is therefore worthwhile to optimize circuits during their creation to the full extent. In this chapter, we describe optimization techniques that minimize a Boolean circuit for an application in MPC protocols with constant rounds.

Optimization Challenge The compilation of an efficient high-level description of a functionality for RAM based architectures does not necessarily translate into an efficient representation as a circuit. For example, in Sect. 4.5 we compare the circuit sizes when compiling a Merge and a Bubble sort algorithm. Merge sort is commonly superior in efficiency when computed on a RAM based architecture, whereas Bubble sort compiles to a significantly smaller circuit representation. Due to these reasons dedicated algorithms have been developed in hardware synthesis (e.g., sorting networks), which are optimized for a circuit based computation. To the best of our knowledge, no generic (and practical) approach is known that is capable of transforming a functionality into a representation that is best suited for circuit synthesis. Therefore, in this chapter *we focus on optimizing the circuit for a given algorithmic representation*. Yet, we remark that a partial solution to the

N. Büscher, S. Katzenbeisser, *Compilation for Secure Multi-party Computation*, SpringerBriefs in Computer Science, https://doi.org/10.1007/978-3-319-67522-0_4

generic problem can be given by using optimized libraries for various tasks (e.g., data structures, sorting, string operations, etc.).

Illustrating the Need for Optimization Optimization is important for the widespread use of MPC. For example, when writing source code for RAM based architectures, developers commonly rely on a few data types that are available, e.g., `unsigned int` or `long`. In contrast, on the circuit level, arbitrary bit-widths can be used. However, it is a tedious programming task to specify precise bit-widths for every variable to achieve minimal circuits. Consequently, optimizing compilers should, for example, identify overly allocated bit-widths on the source code level and adjust them on the gate-level accordingly. Without advanced optimization techniques, the developer is required to be very familiar with circuit synthesis for MPC and compiler internals to write code that compiles into an efficient circuit and thus, efficient application.

To further illustrate the need for optimization in circuit compilation, we study an example code snippet given in Listing 4.1. The main part of the code begins in Line 7, where an input variable is declared that is only instantiated during protocol evaluation. Hence, its actual value is unknown during compile time. Next, a variable `t` is declared and initialized with a constant value of 43,210. Then, a helper function, declared in Line 1, is called that checks whether the given argument is an odd number. Depending on the result of the helper function, `t` will be incremented by one in Line 10.

```
1  int is_odd(int val) {
2    return ((val & 1) == 1);
3  }
4
5  int main() {
6    [...]
7    int INPUT_A_x;
8    int t = 43210;
9    if(is_odd(INPUT_A_x) {
10     t = t + 1;
11   }
12   [...]
13 }
```

Listing 4.1 Code example to illustrate the need for optimization

A naive translation of this code into a Boolean circuit leads to a circuit consisting of four building blocks, namely, a logical AND, an integer equality check, an integer addition as well as a conditional integer assignment. Figure 4.1 illustrates the circuit that is generated by such a direct translation. When using the best known building blocks (these are described in Sect. 4.3) and assuming a standard integer bit-width of 32 bits, 31 or 32 non-linear gates are required for each building block. This results in a total circuit size of $s^{nX} = 2 \cdot 31 + 2 \cdot 32 = 126$ non-linear gates.

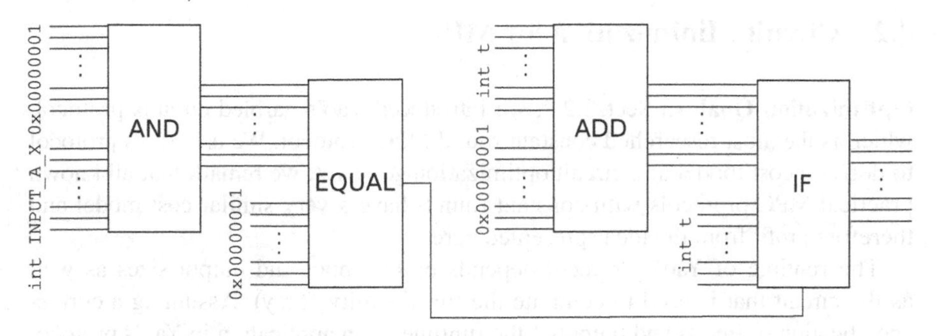

Fig. 4.1 Circuit after naïve translation from source code in Listing 4.1 to three building blocks of bit-width 32

However, an optimizing compiler could detect that the comparison is only a single bit comparison, whose result is equal to the least significant bit of `INPUT_A_x`. Hence, no gate is required for the comparison. Moreover, as variable `t` is initialized by an even constant, the addition in Line 10 can be folded into an assignment of the least significant bit, which leads to a circuit that consists only of a single wire and that does not even contain a single gate, namely:

```
t@0 = INPUT_A_x@0
```

Here we use the notation @*n* to symbolize the accesses to the *n*-th bit, with @0 being the least significant bit. We observe a significant difference in this simple example between naïve translation and optimized compilation. Such a size reduction directly relates to an improvement in the protocol's runtime, as the (amortized) computational cost of an evaluating MPC protocol scales linear with the circuit size.

Therefore, in this chapter we present the optimization techniques implemented in our compiler CBMC-GC that allow the compilation of efficient circuits for MPC protocols with constant rounds, i.e., circuits with a minimal number of non-linear gates. With these techniques, we are able to offer a high level of abstraction from both MPC and circuits by compiling from standard ANSI-C source code. For the creation of optimized circuits, we first provide an overview of optimized building blocks and then present a fixed-point algorithm that combines techniques from logic minimization, such as constant propagation, SAT sweeping, and rewrite patterns, to compile circuits that are significantly smaller than those, generated by related work at the time of writing.

Chapter Outline We first give an overview on the circuit optimization process in CBMC-GC in Sect. 4.2. Then, in Sect. 4.3 we describe size-optimized building blocks for MPC, before describing multiple gate-level optimization techniques in Sect. 4.4. Finally, the effectiveness of these optimization techniques is studied in Sect. 4.5

4.2 Circuit Minimization for MPC

Optimization Goal In Sect. 2.2.2, we introduced Yao's garbled circuits protocol, which is the most researched constant round MPC protocol. We use Yao's protocol to derive a cost model and circuit optimization goal, yet, we remark that all known practical MPC protocols with constant rounds have a very similar cost model and therefore profit from the ideas presented here.

The runtime of Yao's protocol depends on the input and output sizes as well as the circuit that is used to compute the functionality $f(x, y)$. Assuming a correct specification of inputs and outputs,[1] the runtime of an application in Yao's protocol can only be improved by changing the circuit representation $C_f(x, y)$ into a more efficient representation $C'_f(x, y)$. The circuit runtime depends on the number of linear and non-linear gates. Only non-linear gates require computation of encryptions and communication between the parties, whereas the linear gates are considered as being for 'free' (see Sect. 2.2.2), because they neither require communication nor cryptographic operations. Thus, the primary goal of circuit optimization for Yao's protocol, and most other known constant round MPC protocols, is to minimize the number of non-linear gates, such as AND. Nevertheless, once the number of non-linear gates is minimized, as a secondary goal the total number of linear (XOR) gates could also be minimized, because in a practical deployment linear gates also generate (albeit very small) computation costs and require memory accesses.

With communication being the most relevant bottleneck for practical MPC, the techniques described in the next sections mainly aim at achieving the primary goal, yet we avoid an unnecessary increase in XOR gates, whenever possible.

Minimizing Strategy Unfortunately, finding a minimal circuit for a given circuit description is known to be Σ_2^P complete [15]. Therefore, in CBMC-GC we follow a heuristic approach. First, during circuit instantiation (see Chap. 3), we use size-minimized building blocks that are described in Sect. 4.3. Second, the instantiated circuit is optimized on the gate-level by using a best-effort fixed-point optimization algorithm that is discussed in detail Sect. 4.4 and added as an additional step to the compilation chain of CBMC-GC. The full compilation chain of CBMC-GC, including optimizations, is shown in Fig. 4.2.

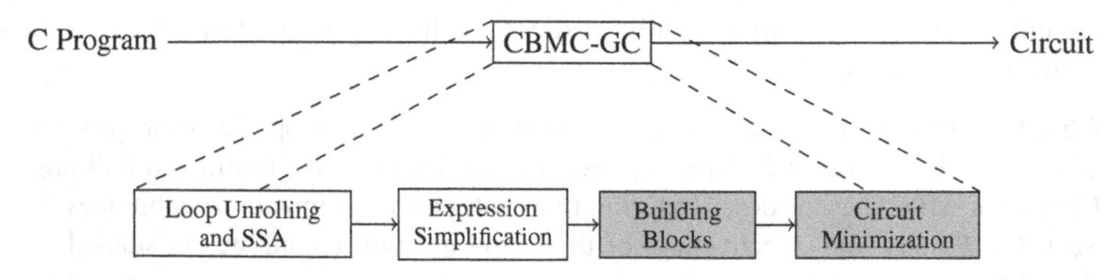

Fig. 4.2 CBMC-GC's abbreviated compilation pipeline with circuit optimization. *Marked in grey* are the parts described in this chapter that lead to size-minimized circuits for MPC

[1]Our optimization methods also detect and display unused input as well as constants output.

4.3 Building Blocks for Boolean Circuit Based MPC

Optimized building blocks are essential for designing complex circuits. They facilitate efficient compilation, as they can be highly optimized once and subsequently instantiated at practically no cost during compilation. In the following paragraphs, we give a comprehensive overview over the currently best known building blocks with a minimal number of non-linear gates for the most common arithmetic and control flow operations.

Adder An n-bit *adder* takes two bit strings x and y of length n, representing two (signed) integers, as input and returns their sum as an output bit string s of length $n+1$. An adder is commonly constructed of smaller building blocks, namely Half-Adders (HA) and Full-Adders (FA). A Half-Adder is a combinatorial circuit that takes two bits A and B and computes their sum $S = A \oplus B$ and carry bit $C_{out} = A \cdot B$. A Full-Adder allows an additional carry-in bit C_{in} as input. The best known constructions [49] for computing the sum bit of a FA is by XOR-ing all inputs

$$S = A \oplus B \oplus C_{in},$$

while the carry-out bit can be computed by

$$C_{out} = (A \oplus C_{in})(B \oplus C_{in}) \oplus C_{in}.$$

Both HA and FA have a non-linear size $s^{nX} = 1$. The standard and best known n-bit adder is the Ripple Carry Adder (RCA) that consists of a successive composition of n FAs. Thus, RCA has a circuit size ${s^{nX}}_{RCA}(n) = n$. We note that, according to the semantics of ANSI-C, an addition is computed as $x + y \mod 2^n$ and no overflow bit is returned, which reduces the circuit size by one non-linear gate to $n-1$.

Subtractor A subtractor for two n bit strings can be implemented with one additional non-linear gate by using the two's complement representation $x - y = x + \bar{y} + 1$. Thus, the addition of negative numbers in the two's complement representation is equivalent to an addition of positive numbers.

Comparator An *equivalence (EQ)* comparator checks whether two input bit strings of length n are equivalent and outputs a single result bit. The comparator can be implemented naïvely by a successive OR composition over pairwise XOR gates that compare single bits. This results in a size of ${s^{nX}}_{EQ}(n) = n - 1$ gates [49]. A *greater-than (GT)* comparator that compares two integers can be implemented with help of a subtractor by observing that $x > y \Leftrightarrow x - y - 1 \geq 0$ and returning the carry out bit, which yields a circuit size of ${s^{nX}}_{GT}(n) = n$.

Multiplier In classic hardware synthesis, a multiplier (MUL) computes the product of two n bit strings x and y, which has a bit-width of $2n$. However, in many programming languages, e.g., ANSI-C, multiplication of unsigned numbers is defined as an $n \rightarrow n$ bit operation by returning $x \cdot y \mod 2^n$. The standard approach for computing an $n \rightarrow 2n$ bit multiplication is often referred to as the "school method". Using a bitwise multiplication and shifted addition, the product is computed as $\sum_{i=0}^{n-1} 2^i(X_i y)$. This approach leads to a circuit requiring n^2 1-bit multiplications and $n-1$ shifted n-bit additions, which in total results in a circuit size of $s^{nX}{}_{MUL}(n) = 2 \cdot n^2 - n$ gates [50]. When compiling a $n \rightarrow n$ bit multiplication with the same method, only half of the one bit multiplications are relevant, leading to a circuit size of $s^{nX}{}_{MUL}(n) = n^2 - n$ gates. The $n \rightarrow n$ multiplication of negative numbers in the two's complement representation can be realized with the same circuit. Alternatively, for a $n \rightarrow 2n$ bit multiplication the Karatsuba-Ofmann multiplication (KMUL) can be used, achieving an asymptotic complexity of $O(n^{\log_2(3)})$. Henecka et al. [40] presented the first adoption for MPC, which was subsequently improved by Demmler et al. [28] by 3% using commercial hardware synthesis tools. Their construction outperforms the school method for larger bit-widths $n \geq 19$.

Multiplexer Control flow operations, e.g., branches and array read accesses, are expressed on the circuit level through multiplexers (MUX). A 2:1 n-bit MUX consists of two input bits strings d^0 and d^1 of length n and a control input bit C. The control input decides which of the two input bit strings is propagated to the output bit string o. For an array read access, a multiplexer with more inputs is required. A 2:1 MUX can be extended to a m:1 MUX that selects between m input strings $d^0, d^1, \ldots, d^m$ using $\lceil \log_2(m) \rceil$ control bits $C_0, C_1, \ldots, C_{\lceil \log(m) \rceil}$ by a tree based composition of 2:1 MUXs. For example, a read access to an array consisting of four bits D_0, D_1, D_2, D_3 and two index bits $i = C_1 C_0$ can be realized as illustrated in Fig. 4.3.

Kolesnikov and Schneider [49] presented a construction of a 2:1 MUX that only requires one single non-linear gate for every pair of input bits by computing the output bit as $O = (D^0 \oplus D^1)C \oplus D^0$. This leads to a circuit size for an n-bit 2:1 MUX of $s^{nX}{}_{MUX}(n) = n$. The circuit size of a tree based m:1 MUX depends on the number of choices m, as well as the bit-width n, yielding

$$s^{nX}{}_{MUX\_tree}(m, n) = (m-1) \cdot s^{nX}{}_{MUX}(n).$$

Demultiplexer Write accesses to an array require a building block, where only the element addressed by a given index is replaced. All other elements should be unchanged. Hence, this resembles very closely the inverse of a multiplexer, referred to as demultiplexer (DEMUX). A 1:m DEMUX has an input index i, an input bit

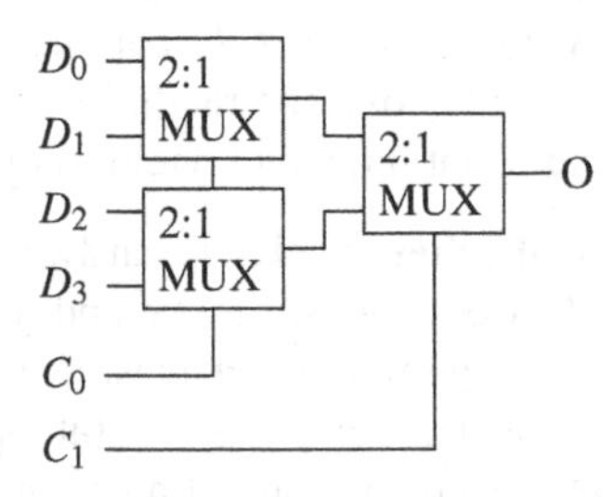

Fig. 4.3 Exemplary array access of four elements compiled into a multiplexer tree

string x, a number m of input bit strings $d_1, d_2, \ldots, d_m$ and outputs m bit strings $d'_1, d'_2, \ldots, d'_m$, where the output d'_j is set to x if $j = i$ and to d_j if otherwise.

A construction for a 1:m DEMUX is given by Malkhi et al. [57]: Each output $d_{out,i}$ is controlled by a multiplexer, which assigns $d'_i \leftarrow x$, if the constant index j is equivalent to the index input bit string i. This construction is similar to a sequence of `if` clauses, i.e., `if(i==0) d[0] = x; if(i==1) d[1] = x;....` This yields a circuit size of $s^{nX}{}_{DEMUX\_EQ}(m,n) = m \cdot (s^{nX}{}_{EQ}(n) + s^{nX}{}_{MUX}(n)) = m \cdot (\lceil \log_2(m) \rceil - 1 + n)$. However, when grouping the equivalence checks in a tree based manner, e.g., equivalence checks for $i = 0001_b$ and $i = 0010_b$ both require to check whether the first two bits are set to zero, the size can be reduced to

$$\begin{aligned}
s^{nX}{}_{DEMUX_tree}(m,n) &= \text{index combinations} + \text{multiplexers} \\
&= \left(m + \frac{m}{2^1} + \frac{m}{2^2} + \cdots + \frac{m}{2^{\lceil \log(m) \rceil - 2}}\right) + (mn) \\
&< 2m + mn = m \cdot (n + 2).
\end{aligned}$$

Divisor A divisor computes the quotient and/or remainder for a division of two binary integer numbers. The standard approach for integer division is known as long division and works similar to the school-method for multiplication. Namely, the divisor is iteratively shifted and subtracted from the remainder, which is initially set to the dividend. Only if the divisor fits into the remainder, which is efficiently decidable by overflow free subtraction, a bit in the quotient is set and the newly computed remainder is used. Thus, a divisor can be built with help of n subtractors and n multiplexers, each of bit-width n, leading to a circuit size of $s^{nX}{}_{SDIV}(n) = 2n^2$. The divisor can be improved by using restoring division [70], which leads to a circuit size of $s^{nX}{}_{RDIV}(n) = n^2 + 2n + 1$.

4.4 Gate-Level Circuit Minimization

As identified in Sect. 4.1, with the compilation being a one-time task, it is very useful to invest computing time in circuit optimization during their compilation. Moreover, as shown, a naïve translation of code into optimized building blocks does not directly lead to a minimal circuit. Therefore, after the instantiation of all building blocks, and thus, construction of the complete circuit, a circuit minimization procedure is run that reduces the number of non-linear gates. CBMC-GC follows an heuristic approach, using a minimization procedure that applies multiple different algorithms to reduce the non-linear circuit size. In the next paragraphs, we first describe the general procedure, before discussing the different components in detail.

CBMC-GC's Minimization Procedure CBMC-GC's minimization routine is illustrated in Fig. 4.4. It begins with the translation of the circuit into an intermediate AND-invert graph (AIG) representation, which is a circuit description that has been

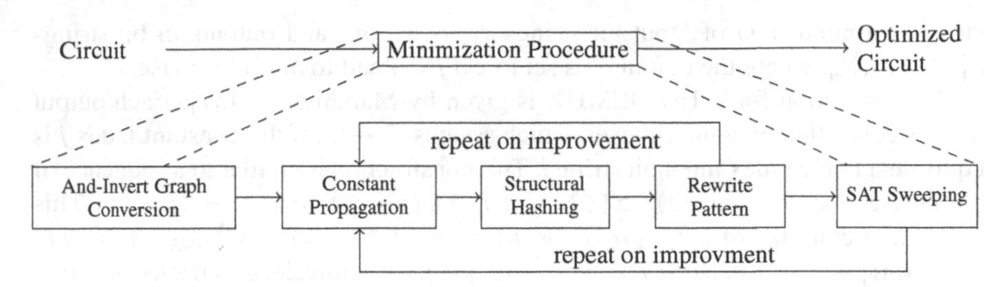

Fig. 4.4 Illustration of CBMC-GC's minimization procedure and its components

shown to allow gate-level optimizations in a very efficient manner. After this AIG translation, a fix-point minimization routine is initiated. The algorithm is run until no further improvements in the circuit size are observed or a user given time bound has been reached. In both cases, the result of the latest iteration is returned.

In every iteration of the algorithm a complete and topological pass over all gates from inputs to outputs is initiated. During this pass, constants (zero or one) are propagated (*constant propagation*), duplicated gate structures are eliminated (*structural hashing*) and small sub-circuits are matched and replaced by hand-optimized sub-circuits (*rewrite patterns*). If any improvement, i.e., reduction in the number of non-linear gates, is observed, a new pass is initiated. If no improvement is observed, a more expensive optimization routine is invoked that detects constant and duplicate gates using a SAT solver (*SAT sweeping*).

AIG, Constant Propagation and Structural Hashing An AND-inverter graph is a representation of a logical functionality using only binary AND gates (nodes) with (inverted) inputs. AIGs have been identified as a very useful representation for circuit minimization, as they allow very efficient graph manipulations, such as addition or merging of nodes. CBMC-GC utilizes the ABC library [7] for AIG handling, which provides state-of-the-art circuit synthesis methods. As a first step in CBMC-GC, input and output wires are created for every input and output variable. Then, during the instantiation of building blocks, the AIG is constructed by substituting every gate type that is different from an AND gate by a Boolean equivalent AIG sub-graph. For example, an XOR gate ($A \oplus B$) can be replaced by the following AIG: $\overline{\overline{V_A \cdot \overline{V_B}} \cdot \overline{\overline{V_A} \cdot V_B}}$, where V_A is the node representing A in the AIG.

Whenever a node is added to the AIG, two optimization techniques are directly applied. First, constant inputs are propagated. Hence, whenever an input to a new node is known as constant, the added node is replaced by an edge. For example, when adding a node V_{new} with inputs from some node V_j and node V_{zero}, which is the node for the constant input zero, then V_{new} will not be added to the AIG. Instead, all edges originating V_{new} will be remapped to V_{zero} instead, because $V_{new} \cdot 0$ is equivalent to 0. Similarly, $V_{new} = V_j \cdot 1$ will be replaced by V_j. Second, structural hashing is applied. Structural hashing [27] is used to detect and remove duplicate sub-graphs, i.e., graphs that compute the same functionality over the same inputs.

Duplicate sub-graphs are also detected when new nodes are added. Hence, when adding a node V_{new} to the AIG, it is checked that no other node exists that uses the same inputs. If such a node V_j is found, it will be replaced by V_j, as described above.

Rewrite Patterns Circuit (and AIG) rewriting is a greedy optimization algorithm used in logic synthesis [59], which was first proposed for hardware verification [11]. A rewrite pattern consists of a template to be matched and a substitute. Both sub-circuits are functionally equivalent. Pattern based rewriting has been shown to be a very effective optimization technique in logic synthesis, as it can be applied with very little computational cost [59]. In CBMC-GC's compilation chain, rewrite patterns are of high importance due to multiple reasons. First, they are responsible for translating the AIG back into a Boolean circuit representation with a small number of non-linear gates. This is necessary, because all linear gates have been replaced by AND gates during the translation in the AIG representation. Second, pattern based rewriting allows for MPC specific optimizations by applying patterns that favor linear gates and reduce the number of non-linear gates. Finally, in CBMC-GC each rewrite pass is also used for constant propagation and structural hashing, as described above. We remark that these techniques are responsible for reducing the bit-width declared on the source-code level to the actual required bit-width by identifying unused or constant gates.

For circuit rewriting all gates are first ordered in topological order by their circuit depth. Subsequently, by iterating over all gates, the patterns are matched against all gates and possible sub-circuits. Whenever a match is found, the sub-circuit becomes a candidate for a replacement. However, the sub-circuit will only be replaced, if the substitution leads to an actual improvement in the circuit size. This is guaranteed for the patterns themselves, which are designed to be minimizing. However, dependencies of intermediate gates, which might be input to other gates, can rule the substitution ineffective. In these cases the sub-circuit will not be replaced. An example is shown in Fig. 4.5, where a sub-circuit matches a template: $(A \cdot B) \oplus (A \cdot C) \rightarrow (A \oplus C) \cdot A$. Yet the detected sub-circuit has intermediate outputs to gates outside the template, which would still need to be computed when rewriting the computation of O.

The outcome and performance of this greedy replacement approach depends not only on the patterns themselves, but also on the order of patterns applied. Therefore, in CBMC-GC, small patterns are matched first, e.g., single gate patterns, as they

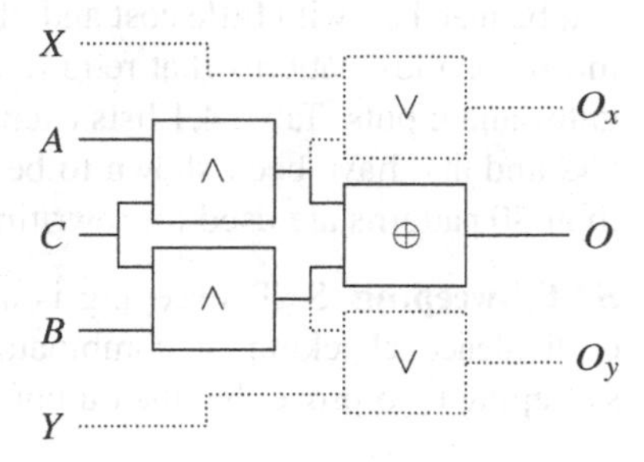

Fig. 4.5 Counter example for circuit rewriting. The sub-circuit indicated by the *solid lines* is a candidate for rewriting. Yet, due to its intermediate outputs, marked with *dotted lines*, a rewriting would actually increase the overall circuit size

Table 4.1 Rewrite patterns. Shown are exemplary rewrite patterns that reduce the circuit size

Search pattern	Substitute	Size reduction (s^{nX}/s)
Propagate pattern		
$\overline{0}$	1	0 / 1
$0 \cdot A$ or $A \cdot 0$	0	1 / 1
$0 + A$ or $A + 0$	A	1 / 1
$0 \oplus A$ or $A \oplus 0$	A	0 / 1
$\overline{1}$	0	0 / 1
$1 \cdot A$ or $A \cdot 1$	A	1 / 1
$1 + A$ or $A + 1$	1	1 / 1
Trivial patterns		
$A \cdot A$	A	1 / 1
$A \cdot \overline{A}$	0	1 / 2
$A + A$	A	1 / 1
$A + \overline{A}$	1	1 / 2
$A \oplus A$	0	0 / 1
$A \oplus \overline{A}$	1	0 / 2
$\overline{\overline{A}}$	A	0 / 2
AND/OR patterns		
$(A \cdot B) \cdot (A \cdot C)$	$A \cdot B \cdot C$	1 / 1
$(A \cdot B) + (A \cdot C)$	$A \cdot (B + C)$	1 / 1
$(A + B) \cdot (A + C)$	$A + (B \cdot C)$	1 / 1
$(A + B) + \overline{A}$	1	2 / 3
$(A \cdot B) \cdot \overline{A}$	0	2 / 3
$(A + B) + (\overline{A} + C)$	1	3 / 4
$(A \cdot B) \cdot (\overline{A} \cdot C)$	0	3 / 4
XOR patterns		
$\overline{A} \oplus \overline{B}$	$A \oplus B$	0 / 2
$(A + B) \oplus (A \cdot B)$	$A \oplus B$	2 / 2
$(A + B) \cdot \overline{(A \cdot B)}$	$A \oplus B$	3 / 3
$\overline{\overline{A} \cdot \overline{B}} \cdot \overline{A \cdot B}$	$A \oplus B$	3 / 6
$(A \cdot (B \oplus (A \cdot C)))$	$A \cdot (B \oplus C)$	1 / 1
$(A \cdot B) \oplus (A \cdot C)$	$(A \oplus C) \cdot A$	1 / 1
$\overline{(A \cdot C) \oplus (B \cdot C)}$	$\overline{(A \oplus B)} \cdot C$	1 / 1

can be matched with little cost and offer guaranteed improvements, before matching more complex patterns that require to compare sub-circuits consisting of multiple gates and inputs. Table 4.1 lists exemplary rewrite patterns that are used in CBMC-GC and that have been shown to be very effective in our evaluation. In total more than 80 patterns are used for rewriting.

SAT Sweeping SAT sweeping is a powerful minimization tool, widely used for equivalence checking of combinatorial circuits [53, 60]. The core idea of SAT sweeping is to prove that the output of a sub-circuit is either constant or equivalent to another sub-circuit (detection of duplicity). In both cases the sub-circuit is unnecessary and can be removed. As common, SAT sweeping is applied in CBMC-GC in a probabilistic manner. A naïve application, which compares every possible combination of sub-circuits, would result in infeasible computational costs. Thus, for efficient equivalence checking, the circuit is first evaluated (simulated) multiple times with different (random) inputs. The gates in every run are then grouped by their output. Gates that always output one or always output zero are presumingly constant, whereas gates that have the same output are presumingly equivalent. This is then proven using the efficient tool of a SAT solver. For this purpose, sub-circuits have to be converted into conjunctive normal form. Due to its high computational cost in comparison with the circuit rewriting, SAT sweeping is only applied if other optimization methods cannot minimize the circuit any further.

4.5 Evaluation

To evaluate the effectiveness of the different presented optimization techniques, we study the circuits generated by CBMC-GC, when compiling the example applications that emerged as standard benchmarks for MPC, described in detail in Sect. 2.3. We first present a comparison of circuits created by the different releases of CBMC-GC, which use an increasing number of optimization techniques described in this chapter. Then, we present a comparison of the circuits generated by CBMC-GC with circuits generated by other state of the art compilers for MPC.

4.5.1 Evaluation of Circuit Minimization Techniques

We abstain from an individual evaluation of the optimization methods presented in the last section. The main reason is that the (de-)activation of a single optimization method has various side-effects that are hard to be controlled and measured in an isolated manner. For example, when using less efficient building blocks, the circuit minimization phase will partly be able to compensate inefficient building blocks and start optimizing these. However, the computation time spent on optimizing the building blocks can consequently not be applied to the remaining parts of the circuit. Similarly, SAT sweeping is highly ineffective without efficient constant propagation. Moreover, some rewrite patterns can become ineffective without the application of other rewrite patterns. Yet, we show that the combination of techniques in CBMC-GC has continuously evolved, such that the circuit sizes compared to earlier releases have significantly been reduced.

Table 4.2 gives a comparison of circuit sizes between the first release of *CBMC-GC v0.8* [42] in 2012, its successor *CBMC-GC v0.9* [32] from 2014, and the *current* version accompanying this book. The initial release of CBMC-GC provided no

Table 4.2 Experimental results: circuit sizes in the number of non-linear gates produced by CBMC-GC v0.8 [42], v0.9 [32], and the current version for various example applications. Moreover, the improvement between the first and the current version of CBMC-GC is shown

Application	CBMC-GC v0.8	CBMC-GC v0.9	CBMC-GC cur	Improvement [%]
Arithmetic 2000	405,640	319,584	253,776	37
Hamming 320 (reg)	6038	924	924	85
Hamming 800 (reg)	15,143	2344	2340	85
Hamming 160 (reg)	30,318	4738	4726	84
Matrix mult. 5×5, int	221,625	148,650	127,255	43
Matrix mult. 8×8, int	907,776	600,768	522,304	42
Median 21, Merge sort	244,720	136,154	61,403	75
Median 31, Merge sort	602,576	348,761	152,823	75
Median 21, Bubble sort	112,800	40,320	10,560	91
Median 31, Bubble sort	349,600	89,280	23,040	93
Scheduling 56, Euclidean	317,544	169,427	113,064	64
Scheduling 56, Manhattan	133,192	88,843	62,188	53

gate-level minimization techniques, yet it contained first optimized building blocks. In CBMC-GC v0.9 the optimization algorithm, described in this chapter, was introduced. All techniques and building blocks have subsequently been improved and adapted for the current version. For example, the building blocks have been revisited and the set of rewrite patterns has been refined, which results in size reductions for all applications.

For comparison, we use the applications introduced in Sect. 2.3. All applications have been compiled from the same source code and optimized with a maximum optimization time of 10 min on a commodity laptop. We observe that the improvement in building blocks is directly visible in the example applications performing random arithmetic operations and a matrix multiplication, which have been improved up to a factor of two, between the first and the current release of CBMC-GC. These two applications purely consist of arithmetic operations that utilize the full bit-width of the used data types, and thus, barely profit from gate-level optimizations. The resulting circuit sizes for the Hamming distance computation show significant improvements when comparing the first and the current release, yet only marginal improvement in comparison to CBMC-GC v0.9. More complex applications, such as Bubble sort based median computation, which involve noticeably more control flow logic, have been improved by more than a factor of ten between the first and the current release of CBMC-GC. Similarly, for the location aware scheduling application, which involves minimizable arithmetic operations as well as control flow logic, we observe an improvement up to a factor of 3 between the first and the current release. In summary, the proposed fix-point optimization routine is very effective to minimize a given circuit.

Table 4.3 Experimental results: comparison between Frigate [61], OblivC [77] and the current version of CBMC-GC. Given are the circuit sizes in the number of non-linear gates when compiling various applications. Marked in bold are significant improvements over the best previous result

Application	Frigate	OblivC	CBMC-GC	Improvement [%]
Biometric matching 128	561,218	560,192	404,419	**27.8**
Biometric matching 256	1,1 M	1,1 M	831,846	**25.7**
Euclidean 5, int	7960	7811	6,235	**20.2**
Euclidean 20, int	25,675	25,001	23,255	7.0
Euclidean 5, fix	26,430	26,307	7,834	**70.2**
Euclidean 20, fix	90,735	90,057	24,559	**12.3**
Float addition	5237	5581	1201	**77.1**
Float multiplication	16,502	14,041	3534	**74.8**
Hamming 160 (reg)	567	899	449	**20.8**
Hamming 1600 (reg)	6546	9269	4738	**27.6**
Hamming 160 (tree)	747	4929	351	**53.0**
Hamming 1600 (tree)	8261	49,569	3859	**53.3**
Hamming 160 (naïve)	1009	4929	541	**46.3**
Hamming 1600 (naïve)	10,282	49,569	6042	**41.2**
Matrix mult. 5×5, int	127,477	127,225	127,225	0
Matrix mult. 5×5, fix	314,000	313,225	183,100	**41.5**
Matrix mult. 5×5, float	2,7 M	2,4 M	626,506	**73.9**
Scheduling 56, Euclidean	133,216	181,071	113,064	**15.1**
Scheduling 56, Manhattan	79,008	77,023	58,800	**23.7**

4.5.2 Compiler Comparison

We compare CBMC-GC with the Frigate [61] and OblivC [77] compilers, which are the most promising candidates for a comparison, as they create circuits with the least number of non-linear gates (at the time of writing) according the compiler analysis by Mood et al. [61]. Even though all compilers use different input languages, we ensure a fair comparison by implementing the functionalities using the same code structure (i.e., functions, loops), data types and bit-widths. Moreover, to present a wider variety of applications, we also investigate more integrated example applications, such the biometric matching application or floating point operations, which are all described in Sect. 2.3. All applications have been compiled with the latest available versions of Frigate and OblivC. For CBMC-GC, we again set a maximum optimization time limit of 10 min on a commodity laptop. The resulting circuit sizes and the improvement of CBMC-GC over the best result from related work are presented in Table 4.3. Circuit sizes above one million (M) gates are rounded to the nearest 100,000.

We observe that the current version of CBMC-GC outperforms related compilers in circuit size for almost all applications. For example, the biometric matching application, or scheduling applications improve by more than 25%. Most significant is the advantage in compiling floating point operations, where a 77% improvement can be observed for the dedicated multiplication operation. A similar improvement is achieved for the computation of the Euclidean distance or matrix multiplication on floating point and fix point values. The operations are dominated by bit wise operations and thus, can be optimized with gate-level optimization when compiled from high-level source code. No improvement is observed for the integer based matrix multiplication. As discussed above, the matrix multiplication compiles into a sequential composition of building blocks, which barely can be improved further.

The Hamming distance computation is well suited to show implementation dependent results and illustrates the challenges of writing source code that compiles into efficient circuits. Here we compare three variants described in more detail in Sect. 2.3. The tree based computation shows the smallest circuit size, even though it is the most inefficient CPU implementation. This is because a tree-based composition allows to apply adders with small bit-widths for a majority of the bit counting. Comparing the compilers, we also observe differences between the implementations. The tree based and naïve implementation are significantly more optimized, i.e., up to a factor of two, in CBMC-GC than in the other compilers. The implementation optimized for register based computation compiles into a circuit that is also smaller in CBMC-GC than in related work, yet only by 20%. This is because the compilation of the naïve bit counting profits significantly from constant propagation, as only a few bits per expression are required on the gate-level. The register optimized implementation maximizes the number of bits used per arithmetic operation, thus, allows only little gate-level optimization. We remark that Frigate and OblivC compile larger applications noticeably faster than CBMC-GC, yet the circuits created by CBMC-GC are up to a factor of four smaller.

Chapter 5
Compiling Parallel Circuits

5.1 Motivation and Overview

At the time of writing, millions of gates can be computed (garbled) in frameworks implementing Yao's garbled circuits on a consumer grade CPU within seconds. Nonetheless, compared with generic computation, Yao's garbled circuits protocol is still multiple orders of magnitude slower. Even worse, an information theoretic lower bound on the number of ciphertexts has been identified by Zahur et al. [78] for gate-by-gate garbling techniques, which makes further simplification of computations unlikely. Observing the ongoing trend towards parallel hardware, e.g., smartphones with many-core architectures on a single chip, the natural questions arises, whether Yao's garbled circuits can be parallelized. In this chapter, we answer this question positively and describe strategies to garble and evaluate circuits in parallel. In particular, we systematically look at two different levels of *compiler assisted parallelization* that have the potential to significantly speed up applications based on secure computation.

Fine-Grained Parallelization (FGP) As the first step, we observe that independent gates, i.e., gates that do not provide input to each other, can be garbled and evaluated in parallel. Therefore, a straight forward parallelization approach is to garble gates in parallel that are located at the same circuit depth, because these are guaranteed to be independent. We refer to this approach as *fine-grained parallelization (FGP)*. We will see that this approach can be efficient for circuits of suitable shape. Nevertheless, the achievable speed-up heavily depends on circuit properties such as the average circuit width, which can be comparably low even for larger functionalities when compiling from a high level language.

Coarse-Grained Parallelization (CGP) To overcome the limitations of FGP for inadequately shaped circuits, we make use of high level circuit descriptions, such as program blocks, to automatically detect larger coherent clusters of gates that can be

N. Büscher, S. Katzenbeisser, *Compilation for Secure Multi-party Computation*, SpringerBriefs in Computer Science, https://doi.org/10.1007/978-3-319-67522-0_5

garbled independently. We refer to this parallelization as *coarse-grained parallelization (CGP)*. We describe how CBMC-GC can be extended to detect concurrency at the source code level to enable the compilation of parallel circuits. Hence, one large circuit is automatically divided into multiple smaller, independently executable circuits. We show that these circuits lead to more scalable and faster execution on parallel hardware. Furthermore, integrating automatic detection of parallel regions into a circuit compiler gives potential users the opportunity to exploit parallelism without detailed knowledge about Boolean circuit based MPC and thus, relieves them of writing parallel circuits.

Chapter Outline Next, we describe the basics of parallel circuits and their parallel evaluation in Yao's garbled circuits in Sect. 5.2. Then we introduce different parallelization approaches as well as a compiler extension to CBMC-GC in Sect. 5.3. Finally, in Sect. 5.4 an evaluation of the presented parallelization approaches in a practical setting alongside example applications is given.

5.2 Parallel Circuit Evaluation

We first discuss the basic notations of sequential and parallel circuit decomposition used throughout this chapter, before describing how Yao's garbled circuits protocol can be parallelized.

Parallel and Sequential Circuit Decomposition In this chapter, we again consider a given functionality $f(x, y)$ with two input bit strings x, y (representing the inputs of the parties) and an output bit string o. Furthermore, we use C_f to denote the circuit that represents functionality f. We refer to a functionality f as *sequentially decomposable* into two *sub-functionalities* f_1 and f_2 iff $f(x, y) = f_2(f_1(x, y), x, y)$.

Moreover, we consider a functionality $f(x, y)$ as parallel decomposable into sub-functionalities $f_1(x, y)$ and $f_2(x, y)$ with non-zero output bit length, if a bit string permutation σ_f exists such that $f(x, y) = \sigma_f(f_1(x, y)||f_2(x, y))$, where $||$ denotes bitwise concatenation.

Thus, functionality f can directly be evaluated by independent evaluation of f_1 and f_2. We observe that the permutation σ_f is only a formal requirement, yet when representing f_1 and f_2 as circuits, the permutation is instantiated by connecting input and output wires at no cost. Furthermore, we note that f_1 and f_2 do not necessarily have to be defined over all bits of x and y. Depending on f they could share none, some, or all input bits. We use the operator $\diamond$ to express a parallel composition of two functionalities through the existence of a permutation σ. Thus, we write $f(x, y) = f_1(x, y) \diamond f_2(x, y)$ if there exists a permutation σ_f such that $f(x, y) = \sigma_f(f_1(x, y)||f_2(x, y))$.

We call a parallelization of f to be *efficient* if the circuit size (i.e., number of gates) of the parallelized functionality is roughly equal to the circuit size of the sequential functionality: $size(C_f) \approx size(C_{f_1}) + size(C_{f_2})$. More precisely, due to the different garbling methods for linear and non-linear gates in Yao's protocol using the

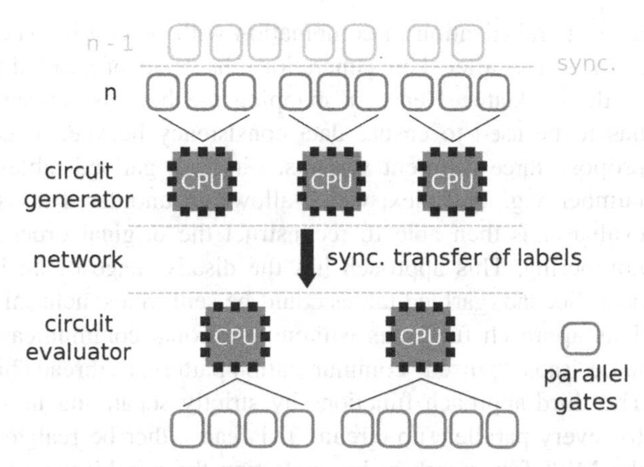

Fig. 5.1 Interaction between a parallel circuit generator and evaluator. The layer n of the presented circuit is garbled and evaluated in parallel. The independent partitions of the circuit can be garbled and evaluated by different threads in any order

free-XOR technique, we only count the number of non-linear gates. Furthermore, we refer to a parallelization as *symmetric* if sub-functionalities have almost equal circuit sizes: $size(C_{f_1}) \approx size(C_{f_2})$.

Finally, we refer to functionalities that can be decomposed into a sequential and a parallel part as *mixed functionalities*. For example the functionality $f(x,y) = f_3(f_1(x,y) \diamond f_2(x,y), x, y)$ can first be decomposed sequentially in f_3 and $f_1 \diamond f_2$, where the latter part can then be further decomposed in f_1 and f_2. Without an explicit formalization, we note that all definitions can be extended from the dual case f_1 and f_2 to the general case $f_1, f_2, \ldots, f_n$.

Parallel Circuit Creation and Evaluation A decomposition of a circuit into sequential and parallel sub-circuits forms a directed acyclic graph (DAG) from inputs to output bits. Hence, sequentially aligned sub-circuits have to be garbled in order, whereas parallel aligned sub-circuits can be garbled in any order using standard garbling techniques, described in Sect. 2.2.2. Parallel sub-circuits can be garbled in any order by one or multiple computing units (threads). This is exemplary illustrated in Fig. 5.1, where three threads are used to garble and two threads to evaluate a circuit. We note that the garbling order has no impact on the security [54]. After every parallel decomposition a synchronization between the different threads is needed to guarantee that all wire labels for the next sequential sub-circuit are computed. Multiple subsequent parallel regions with possibly different degrees of parallelism can be garbled, when ensuring synchronization in between.

The circuit evaluation can be parallelized in the same manner. Sequential sub-circuits are computed sequentially, parallel sub-circuits are computed in parallel by different threads. After every parallelization a thread synchronization is required to ensure data consistency.

For efficiency reasons, implementations of Yao's garbled circuits commonly do not need to store the gate id next to the computed garbled table, as their garbling and evaluation order is deterministic. Hence, given a list of garbled tables, the garbled table associated with each gate can be identified by the position in the list. When using parallelization in combination with pipe-lining (i.e., garbled tables are sent immediately after their generation) the order of garbled tables could be unknown to the evaluator. Hence, a mapping mechanism between gate and garbled table has to be used to ensure data consistency between generator and evaluator. We propose three different variants. First, all garbled tables can be enriched with a number, e.g., an index, which allows an unordered transfer to the evaluator. The evaluator is then able to reconstruct the original order based on the introduced numbering. This approach has the disadvantage of an increased communication cost. Second, garbled tables could be sent in a synchronized and predefined order. This approach functions without additional communication, yet could lead to an undesirable "pulsed" communication pattern, as threads have to wait for each other. The third approach functions by strictly separating the communication channels for every parallel sub-circuit. This can either be realized by multiplexing within the MPC framework or by exploiting the capabilities of the underlying operating system. Due to the aforementioned reasons, the results described in Sect. 5.4 are based on an implementation using the latter approach.

5.3 Compiler Assisted Parallelization Heuristics

To exploit parallelism in Yao's protocol, groups of gates that can be garbled independently need to be identified. As described, independent gates can be garbled and evaluated in parallel. However, detecting independent, similar sized groups of gates is known as the NP-hard graph partitioning problem [58]. The common approach to circumvent the expensive search for an optimal solution is to use heuristics. In the following paragraphs we study a fine- and a coarse-grained heuristic, where the first operates on the gate level and the latter on the source code level.

5.3.1 Fine-Grained Parallelization (FGP)

A first heuristic that decomposes a circuit into independent parts is the *fine-grained* gate level approach. Similar to the evaluation of a standard Boolean circuit, gates in garbled circuits are processed in topological execution order. Gates provide input to other gates and hence, can be ordered by the circuit level (depth) when all their inputs are ready or the level when their output is required for succeeding gates. Consequently, gates on the same level can be garbled in parallel [2, 44]. Thus, a circuit is sequentially decomposable into different levels and each level is further decomposable in parallel with a granularity up to the number of gates in each level. Figure 5.2 illustrates fine-grained decomposition of a circuit into three levels L1, L2 and L3.

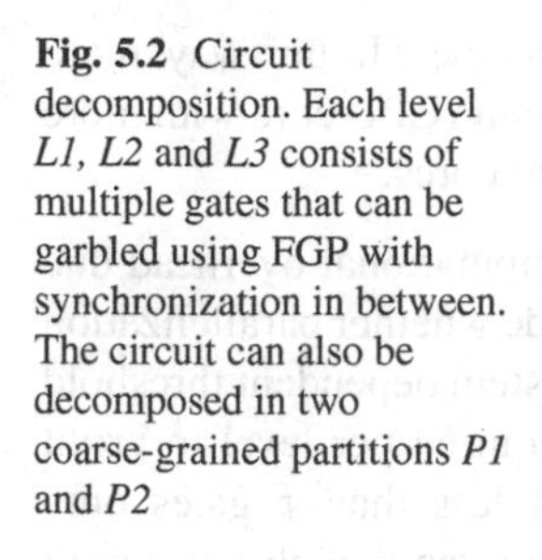

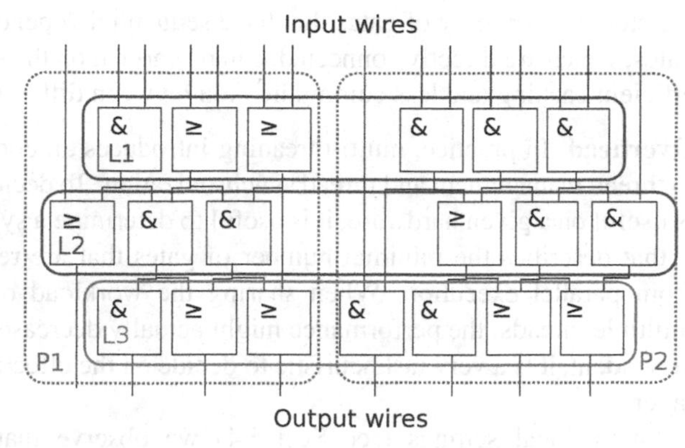

Fig. 5.2 Circuit decomposition. Each level *L1, L2* and *L3* consists of multiple gates that can be garbled using FGP with synchronization in between. The circuit can also be decomposed in two coarse-grained partitions *P1* and *P2*

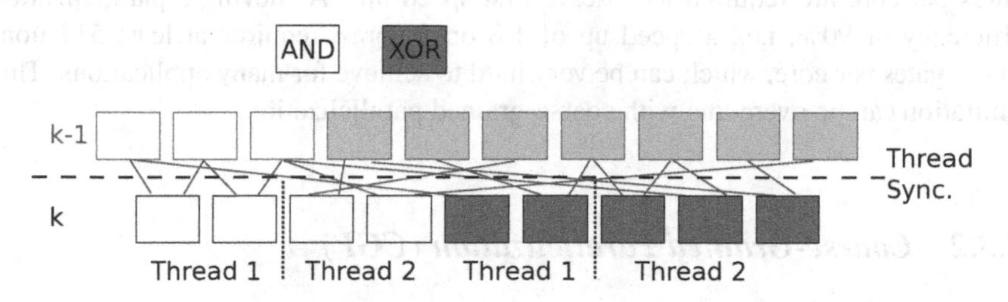

Fig. 5.3 Illustration of the fine grained parallelization approach for one party. Level *k* is distributed symmetrically on two threads. Each thread is tasked to garble two non-linear gates. In-between two levels, a thread synchronization is needed to ensure data availability

Compiler-Assisted FGP in Yao's Garbled Circuits With millions of gates garbled and evaluated during the protocol execution, it is useful to identify and annotate the circuit levels already during compilation to achieve an efficient distribution of gates onto threads during protocol runtime. Furthermore, FGP can be improved by considering linear and non-linear gates independently (because they have very different workloads) when distributing them onto all threads, as this enables a more symmetric workload distribution among multiple threads. Consequently, the computation will not be stalled by threads waiting for other threads. Therefore, each thread gets assigned a similar number of linear and non-linear gates to garble per circuit level. This is exemplary illustrated in Fig. 5.3, where two threads share the task of garbling a circuit layer with a similar workload.

We extended the CBMC-GC compiler with the capability to mark levels and to strictly separate linear from non-linear gates within each level. This information is stored in the circuit description that is then interpreted in a protocol implementation.

Finally, we remark that when using Non-Uniform Memory Access (NUMA) hardware, e.g., multiple cores on a CPU that differentiate between local and shared storage or multiple CPUs on different sockets, the efficiency of FGP can further be improved by fine-tuning the distribution of gates onto threads by maximizing the storage proximity of gates that have sequential dependencies. In this way, more gates, which are directly connected, can be garbled on the same CPU core with more efficient caching and less communication between different cores.

Overhead In practice, multi-threading introduces an computational overhead due to thread management and thread synchronization. To decide whether parallelization is useful on a given hardware, it is useful to determine a system dependent threshold τ that describes the minimal number of gates that are required per level to profit from parallel execution. When sharing the workload of less than τ gates onto multiple threads, the performance might actually decrease. Even though τ is circuit dependent, it is a very fast heuristic to decide on the effectiveness of FGP per circuit layer.

In practical settings (see Sect. 5.4) we observe that at least $\sim$8 non-linear gates per core are required to observe first speed-ups. Achieving a parallelization efficiency of 90%, i.e., a speed up of 1.8 on 2 cores, requires at least 512 non-linear gates per core, which can be very hard to achieve for many applications. This limitation can be overcome with coarse-grained parallelization.

5.3.2 *Coarse-Grained Parallelization (CGP)*

Another useful heuristic to partition a circuit is to use high level functionality descriptions. Given a circuit description in a high level language, parallelizable regions of the code can be identified using code analysis techniques. These detected code regions allow a parallel code decomposition; independent code parts can then be compiled into *sub-circuits* that are guaranteed to be independent of each other and therefore can be garbled in parallel. We refer to this parallelization scheme as *coarse-grained parallelization (CGP)*. Figure 5.2 illustrates such a decomposition for an exemplary circuit in two coarse-grained partitions P1 and P2. In the following paragraphs, we introduce a compiler extension for CBMC-GC that automatically produces coarse-grained parallel circuits. Furthermore, we note that FGP and CGP can be combined by utilizing FGP within all coarse partitions.

Compiler Extension for CGP in Yao's Garbled Circuits Our parallel circuit compiler extension *ParCC* extends CBMC-GC to compile circuits with coarse-grained parallelism. Its core functionality is to identify data parallelism in loops and to decompose the source code in a pre-processing step before its translation onto the circuit level. Conceptually, ParCC detects parallelism within ANSI-C code carrying CBMC-GC's I/O annotations and compiles a *global circuit* that is interrupted by one or multiple *sub-circuits* for every parallel code region. The global circuit and all sub-circuits are interconnected by *inner* input and output wires. These are not exposed as inputs or outputs to the MPC protocol, but allow the recombination of the complete and parallel executable circuit. If a parallel region follows the single-instruction-multiple-data paradigm (SIMD), it is sufficient to compile only one sub-circuit to save compilation time as well as storage cost.

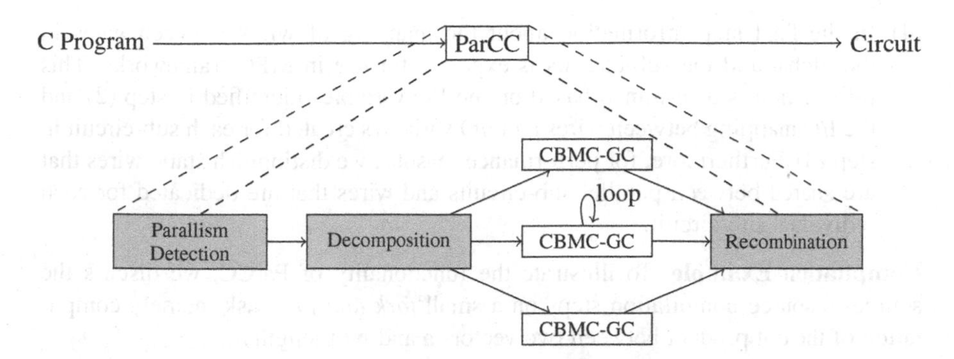

Fig. 5.4 ParCC's compilation pipeline. First, parallelism in the input source is detected. Then, the code is decomposed in parallel and sequential parts. Each part is compiled individually with an unmodified variant of CBMC-GC. Finally, the resulting sub-circuits are recombined into a single circuit with annotated parallelism

Our implementation is built on top of state of the art source code parallelization tools to detect parallelism and to perform code transformations. Namely, the parallellization framework *Pips* [14] is used to detect and to annotate parallelism with the polyhedral loop optimizer *POCC* [68]. The source-to-source transformation of the annotated code, i.e., the program decomposition described below, is realized with the help of Pips, as well the static source code analysis toolkit *Frama-C* [23]. The complete compilation process, as illustrated in Fig. 5.4, consists of four different steps:

(1) In the first step, parallelism in C code is detected by one of the algorithms provided by Pips and annotated using the *OpenMP* notation [25]. OpenMP is the de-facto application programming interface for shared memory multiprocessing programming in C.
(2) The annotated C code is parsed by ParCC in the second step. According to the identified and annotated parallelism, the source code is decomposed using source-to-source techniques into a *global* sequentially executable part, which is interrupted by one or multiple parallel executable *sub*-parts. This is realized by exporting each loop body into an individual function, and providing an interface between the original code location and the newly introduced function in form of CBMC-GC I/O variables. Additionally, OpenMP reduction statements, e.g, `sum`, are replaced with a code template and are added to the global part that is later compiled into an according circuit. Reduction functions require a separate treatment, as they are not embarrassingly parallel. Furthermore, information about the degree of detected parallelism as well as the interface between the global and sub-parts is extracted for later compilation steps.
(3) Given the decomposed source code, the different parts are compiled independently with CBMC-GC. Hence, one global and multiple independent sub-circuits are created.

(4) In the final step information about the mapping of wires between gates in the global and the sub-circuits is exported for use in MPC frameworks. This information is determined based on the I/O variables identified in step (2) and the I/O mapping between wires and I/O variables created for each sub-circuit in step (3). Furthermore, for performance reasons, we distinguish static wires that are shared between parallel sub-circuits and wires that are dedicated for each individual sub-circuit.

Compilation Example To illustrate the functionality of ParCC, we discuss the source-to-source compilation steps on a small *fork and join* task, namely computation of the dot product between two vectors **a** and **b** of length n:

$$r = \mathbf{a} \cdot \mathbf{b} = a_0 \cdot b_0 + \cdots + a_{n-1} \cdot b_{n-1}.$$

The source code of the function `dot_product()` is presented in Listing 5.1.

```
1 void dot_product() {
2   int INPUT_A_a[100], INPUT_B_b[100];
3   int res = 0;
4   for(i = 0; i < 100; i++)
5     res += INPUT_A_a[i] * INPUT_B_b[i];
6   int OUTPUT_res = res;
7 }
```

Listing 5.1 Dot vector product written in C with CBMC-GC input/output notation

In this example code, two parties provide input for one vector in form of constant length integer arrays (Line 2). A loop iterates pairwise over all array elements (Line 4), multiplies the elements and aggregates the result. In the first compilation step, Pips detects the loop parallelism and annotates this parallel region accordingly. The annotated code is printed in Listing 5.2, with the OpenMP annotation added in Line 2.

```
1 [...]
2 #pragma omp parallel for reduction(+:res)
3 for(i = 0; i <= 99; i++) {
4   res += INPUT_A_a[i] * INPUT_B_b[i];
5 }
6 [...]
```

Listing 5.2 Annotation of parallelism as detected and added by the Pips framework after the first compilation step of the dot vector product example

ParCC parses the annotations in the second compilation step and exports the loop body in an own function named `loop0()`, as it is the first loop encountered during compilation. The exported function is printed in Listing 5.3.

```
1 void loop0(int INPUT_A_0, int INPUT_A_1, int OUTPUT_return)
2 {
3   OUTPUT_LOOP0_return = INPUT_A_0 * INPUT_A_1;
4 }
```

Listing 5.3 Exported sub-function with CBMC-GC input-output notation

The functions expects input, in this example two integer variables, according to the notation of CBMC-GC. The result is returned in form of an (inner) output variable. Note, that during the protocol execution all inner wires are not assigned to any party, instead they connect gates in the global circuit and sub-circuits. Yet, to keep compatibility with CBMC-GC a concrete assignment for the party P_A is specified. The later exported mapping information is used to distinguish between inner wires and actual input wires of both parties. Moreover, in the same step, the global function `dot_product ()`, printed in Listing 5.4, is transformed by ParCC to replace the loop by an unrolled version of itself.

```
1  void dot_product() {
2    int INPUT_A_a[100], INPUT_B_b[100];
3    int res = 0;
4    int OUTPUT_LOOP0_a[100];
5    int OUTPUT_LOOP0_b[100];
6    int i;
7    for(i = 0; i <= 99; i++) {
8      OUTPUT_LOOP0_a[i] = INPUT_A_a[i];
9      OUTPUT_LOOP0_b[i] = INPUT_B_b[i];
10   }
11   int INPUT_A_LOOP0_res[100];
12   for(i = 0; i <= 99; i++)
13     res += INPUT_A_LOOP0_res[i];
14   int OUTPUT_res = res;
15 }
```

Listing 5.4 Rewritten function `dot_product()`. The loop has been replaced by inner input/output variables (marked with `LOOP0`)

For this purpose, the two arrays `INPUT_A_a` and `INPUT_B_b`, which are iterated over in the original loop, are now exposed as inner output variables to create the mapping between the global circuit and sub-circuits. Therefore, from Line 4 to Line 10 ParCC added two new output arrays using CBMC-GC's I/O notation that are assigned to the two arrays. Furthermore, an inner input array for the results of the exported sub-functionality is introduced in Line 11. Finally, the reduction statement is substituted by synthesized additions over all intermediate results in Line 13. Multiple parallel parts in a given source code are exported and handled individually by independent decomposition of each part.

5.4 Evaluation of Parallelization in Yao's Garbled Circuits

We begin by introducing a parallel Yao's garbled circuits framework named *UltraSFE* and benchmark its performance on a single core in Sect. 5.4.1. The applications and their circuit descriptions used for benchmarking are described in Sect. 5.4.2. We evaluate the offline garbling performance of the proposed parallelization techniques in Sect. 5.4.3, before evaluating the promising CGP in an online setting in Sect. 5.4.4.

5.4.1 UltraSFE

UltraSFE is a framework for Yao's garbled circuits that implements FGP and CGP. To realize efficient parallelization, all data structures, the memory layout, and the memory footprint are all optimized with the purpose of parallelization in mind. All these implementation optimizations are of importance, due to the specific resource requirements of Yao's garbled circuits. When garbling millions of gates per second, memory read and write accesses quickly become a bottleneck. Wire labels in the size of hundred(s) of megabytes per second have to be fetched and written from and to memory in a unaligned manner. Therefore, a reduction of the overall memory footprint leads to better caching behavior, which becomes even more important when using multi core architectures.

UltraSFE is written in C++ using SSE4, OpenMP and Pthreads to realize multi-core parallelization. Conceptually, UltraSFE implements the fixed-key garbling scheme *JustGarble* [5] and use ideas from the Java based memory efficient *ME_SFE* framework [41], which itself is based on the *FastGC* framework [43]. Oblivious transfers are realized with the help of the highly efficient and parallelized OTExtension library written by Asharov et al. [1]. Moreover, UltraSFE adopts the best known techniques and optimizations for Yao's protocol. This includes pipe-lining, point-and-permute, garbled row reduction, free-XOR and the half-gate approach [43, 49, 57, 67, 78].

Framework Comparison To illustrate the practical performance gains through parallelization schemes in a fair manner, it is necessary to compare the results to a highly optimized single core implementation. To illustrate that UltraSFE is suited to evaluate the scalability of different parallelization approaches, we present a comparison of its garbling performance with other state-of-the-art (at the time of writing) frameworks in Table 5.1. Namely, we compare the single core garbling speed of UltraSFE, which is practically identical to the performance of JustGarble (*JG*) by Bellare et al. [5], with the parallel frameworks by Barni et al. (*BCPU*) [2], Husted et al. (*HCPU*) [44], Kreuter et al. (*KSS*) [51], and *GraphSC* by Nayak et al. [65]. Note, these results are compared in the *offline* setting, i.e., truth tables are written to memory, rather then sent to the evaluator. This is because circuit garbling is the most cost intensive part of Yao's protocol and therefore the most

Table 5.1 Circuit garbling. Single core garbling speed comparison of different frameworks on circuits with more than five million gates. Metrics are the number of *non-linear gates per second* that can be garbled on a single core in millions (M) and CPU *clocks per gate*. All results have been observed on the Intel processor specified in row *architecture*. Note, for HCPU [44] only circuit evaluation times have been reported on the CPU, the garbling speed can be assumed to be lower, as it requires four times the number of encryptions

	BCPU [2]	HCPU [44]	KSS [51]	GraphSC [65]	JG [5]/UltraSFE
Gates per second	0.11M	<0.25M	0.1M	0.58M	8.3M
Clock cycles per gate	>3500	–	>6500 [5]	> 1200	$\sim$ 110
Architecture	E5-2609	E5-2620	i7-970	E5-2666 v3	E5-2680 v2

interesting when comparing the performance of different frameworks. The previous parallelization efforts HCPU and BCPU actually abstained from implementing an *online* version of Yao's protocol that supports pipe-lining. As metrics we use garbled gates per second and the average number of CPU clock cycles per gate, as proposed in [5] for benchmarking purposes. The numbers are taken from the cited publications and if not given, the clock cycles per gate results are calculated based on the CPU specifications. Even when considering theses numbers only as rough estimates, due to the different CPU types, we observe that UltraSFE performs approximately 1–2 orders of magnitude faster than existing parallelizations of Yao's protocol. This is mostly due to the efficient fixed-key garbling scheme using the AES-NI hardware extension and a carefully optimized implementation using SSE4. Summarizing, UltraSFE shows competitive garbling performance on a single core and hence, is a very promising candidate to study the effectiveness of parallelization.

5.4.2 Evaluation Methodology

To evaluate the different parallelization approaches we use three example applications that have been used to benchmark and compare the performance of Yao's garbled circuits in the past. A detailed description of these can be found in Sect. 2.3. The benchmarked applications and the chosen configurations are:

- *Biometric matching (BioMatch).* In BioMatch one party matches a biometric sample against the other's party database of biometric templates. In the evaluation we use the squared Euclidean distance as distance function between two samples, a database of size $n = 512$, the degree (number of features) is set to $d = 4$, and a integer bit-width of $b = 64$ bit. This values have been used in previous works on MPC [29, 48].
- *Parallel Modular exponentiation (MExp).* MExp can be used for blind signatures and its parallel version in online signing services. We study a circuit with $k = 32$ parallel instances of modular exponentiation and an integer bit-width of $b = 32$ bit.

Table 5.2 Circuit properties. Presented are the code size, the overall circuit size in the number of gates, the fraction of non-linear gates that determine the majority of computing costs, the number of input bits as well as the sequential offline garbling time with UltraSFE

	BioMatch	MExp	MVMul
Code size	22 LOC	28 LOC	10 LOC
Circuit size	66M	21.5M	3.3M
Fraction of non-linear gates	25%	41%	37%
# Input bits P_A/P_B	131K/256	1K/1K	17K/1K
Offline garbling time	2.07 s	1.136 s	0.154 s

- *Matrix-vector multiplication (MVMul).* MVMul is a building block for many privacy-preserving applications. We parametrize this task according to the size of the matrix $m \times k = 16 \times 16$ and vector $k = 16$, as well the integer bit-width of each element $b = 64$ bit.

Circuit Creation All circuits are compiled twice with CBMC-GC using textbook C implementations, once with ParCC enabled and once without. The time limit for the circuit minimization through CBMC-GC is set to 10 min. The resulting circuits and their properties are shown in Table 5.2. The BioMatch circuit is the largest circuit and has the most input bits. The MVMul circuit garbles in a fraction of a second and thus, is suitable to evaluate the performance of parallelization on smaller circuits. The MExp circuit shows a large circuit complexity in comparison to the number of input bits. Even so not shown here, we note that the sequential (CBMC-GC) and parallel (ParCC) circuits slightly differ in the overall number of non-linear gates due to the circuit minimization techniques of CBMC-GC, which profit from decomposition.

Environment As testing environment we used Amazon EC2 cloud instances. These provide a scalable number of CPUs and can be deployed at different sites around the globe. If not state otherwise, for all experiments instances of type `c3.8xlarge` have been used. These instances report 16 physical cores on two sockets with CPUs of type Intel Xeon E5-2680v2, and are equipped with a 10 Gbps ethernet connection. A fresh installation of Ubuntu 14.04 was used to ensure as little background noise as possible. UltraSFE was compiled with `gcc 4.8 -O2` and `numactl` was utilized when benchmarking with only a fraction of the available CPUs. Numactl allows memory, core and socket binding of processes. Results have been averaged over ten executions.

Methodology Circuit garbling is the most expensive task in Yao's protocol. Therefore, we begin by evaluating FGP and CGP for circuit garbling independent of other parts of Yao's protocol. This allows an isolated evaluation of the computational performance gains through parallelization. Following the offline circuit garbling phase is an evaluation of Yao's full protocol in an online LAN setting. This evaluation also considers the bandwidth requirements of Yao's protocol.

5.4.3 Circuit Garbling (Offline)

We begin the evaluation of FGP and CGP by studying the independent task of circuit garbling, which can be executed by the generator offline in a pre-processing phase. In practice the efficiency of any parallelization is driven by the ratio between computational workload per thread and synchronization between threads. When garbling a circuit with FGP, the workload is bound by the width of each level, when garbling with CGP the workload is bound by the size of parallel partitions. Both parameters are circuit and hence, application dependent.

Thread Utilization To get a better insight, we first empirically evaluate the possible efficiency gain for different sized workloads, independent of any application. This also allows to observe a system dependent threshold τ, introduced in Sect. 5.3.1, which describes the minimal number of gates required per thread to profit from parallelization. Therefore, the following experiment was run in the environment previously described. For level widths $w \in \{2^4, 2^5, \ldots, 2^{10}\}$ we created artificial circuits of depth $d = 1000$. The width is kept homogeneous in all levels. Furthermore, the wiring between gates on different levels is randomized and only non-linear gates are used. Each circuit is garbled using FGP and we measured the parallelization efficiency, which is the speed-up over the single core performance divided by the number of cores, when computing with different numbers of threads. The results are illustrated in Fig. 5.5.

The experiment shows that on the tested system $\tau \approx 8$ non-linear gates per thread are sufficient to observe first performance gains through parallelization when using CPU cores that are located on the same socket. To achieve an efficiency of 90% approximately 512 non-linear gates per thread are required. Investigating the results for 16 parallel threads, we observe that a significantly larger workload per thread (at least one order of magnitude) is required to achieve the same efficiency, as communication latency between multiple sockets has to be overcome.

Example Applications We evaluate the speed-up of circuit garbling when using FGP and CGP for the three applications BioMatch, MExp and MVMul compiled with ParCC disabled (FGP) and enabled (CGP). The speed-up is calculated in relation to the single core garbling performance given in Table 5.2. The results, which have been observed for a security level of $k = 128$ bit, are presented in Fig. 5.6. Discussing the results for FGP, we observe that all applications profit from parallelization. BioMatch and MExp show very limited scalability, whereas the MVMul circuit can be executed with a speed-up of 7.5 on 16 cores. Analyzing the performance of CGP, we observe that all applications achieve practically ideal parallelization when using up to four threads. In contrast to the FGP approach, scalability with high efficiency is observable with up to eight threads. When scaling to 16 threads (two sockets), significant further speed-ups are noticeable in the MExp and MVMul experiments, with a total throughput of more than 100M non-linear gates per second.

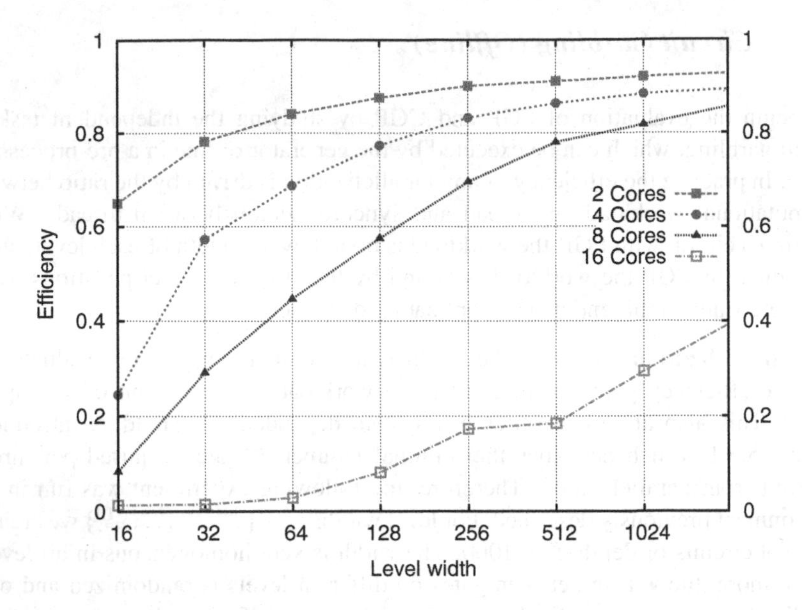

Fig. 5.5 Thread utilization experiment. Shown is the efficiency of FGP for different circuit level widths. A larger width increases the efficiency of parallelization. The gap between 8 (one socket) and 16 cores (two sockets) is due to the communication latency between two sockets on the used hardware

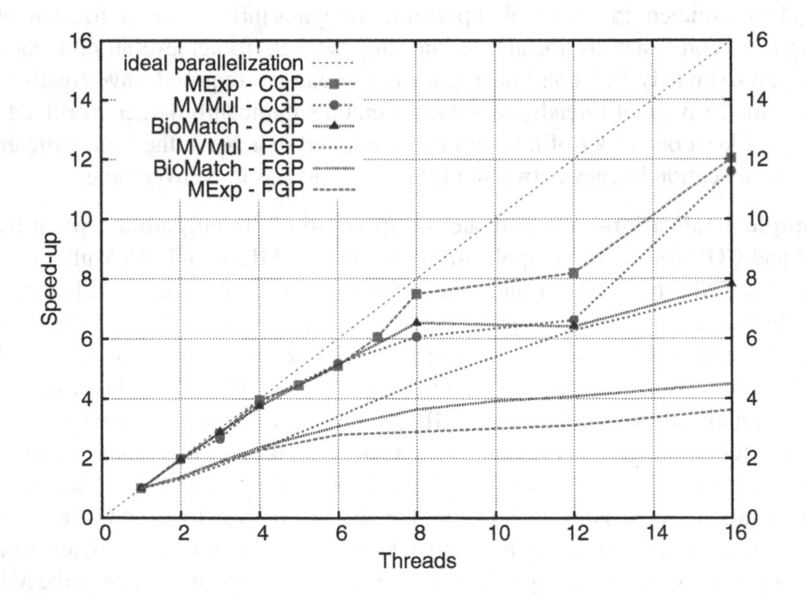

Fig. 5.6 Circuit garbling. The speed-up of circuit garbling for all three applications when using the FGP, CGP and different numbers of computing threads. CGP significantly outperforms FGP for all applications

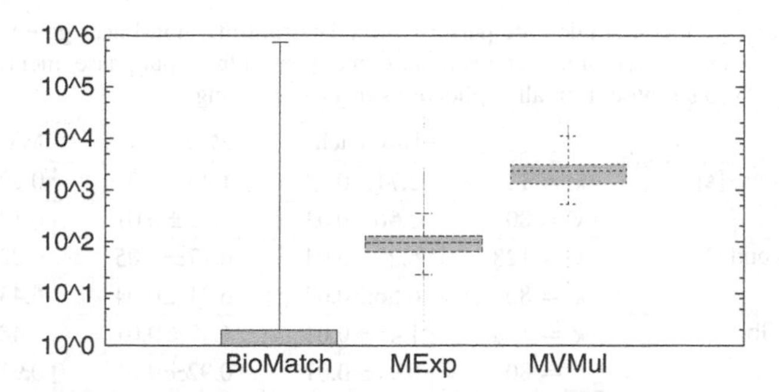

Fig. 5.7 Circuit width distribution. Shown is the distribution of non-linear gates per level when compiling the BioMatch, MExp and MVMul application with CBMC-GC

In summary, for all presented applications the CGP approach significantly outperforms the FGP approach regarding scalability and efficiency due to its coarser granularity, which implies a better thread utilization.

Circuit Width Analysis The limited scalability of FGP is explainable when investigating the different circuit properties. In Fig. 5.7 the distribution of level widths for all circuits are shown when compiled with CBMC-GC.

For the MVMul application, the CBMC-GC compiler produces a circuit with a median level width of 2352 non-linear gates per level, whereas the BioMatch and MExp circuits only show a median width below 100 non-linear gates per level. The major reason for small circuit widths in comparison to the overall circuit size is that high level MPC compilers such as CBMC-GC have been developed with a focus on minimizing the number of non-linear gates. Minimizing the circuit depth or maximizing the median circuit width barely influence the sequential runtime of Yao's protocol and are therefore not addressed in the first place. These observations are also a motivation for the next chapter, where we study the depth minimization of circuits. Looking at the building blocks that are used in CBMC-GC, we observe that arithmetic blocks (e.g., adder, multiplier) show a linear increase in the average circuit width when increasing the input size. Integer comparisons have a constant circuit width for any input bit size. However, multiplexers, as used for dynamic array accesses and for `if`-statements, show a circuit width that is independent (constant) of the number of choices. Thus, a 2-1 multiplexer and a n-1 multiplexer are compiled to circuits with similar sized levels, yet with different circuit depths. Based on these insights we deduce, that the MVMul circuit shows a significantly larger median circuit width, because of the absence of any dynamic array access, conditionals or comparisons. This is not the case with the BioMatch and MExp applications. Considering that every insufficient saturation of threads leads to an efficiency loss of parallelization, we conclude that scalability of FGP is not guaranteed when increasing input sizes.

Table 5.3 Yao's protocol, single-core performance. The runtime, non-linear gate throughput in million *gates per second*, required *bandwidth* and time spent in the input phase, including the OTs when executing Yao's protocol for all applications in a LAN setting

		BioMatch	MExp	MVMul
Protocol runtime [s]	$\kappa = 128$	2.71±0.02	1.43±0.01	0.20±0.00
	$\kappa = 80$	2.56±0.03	1.42±0.01	0.19±0.00
Gates per second [M]	$\kappa = 128$	6.23±0.04	6.17±0.05	6.22±0.00
	$\kappa = 80$	6.56±0.07	6.21±0.04	6.43±0.00
Bandwidth [Gbps]	$\kappa = 128$	1.48±0.01	1.47±0.01	1.48±0.00
	$\kappa = 80$	0.97±0.01	0.92±0.01	0.95±0.00
Input phase duration [s]	$\kappa = 128$	<0.02	<0.01	< 0.01
	$\kappa = 80$	<0.02	<0.01	< 0.01

5.4.4 Full Protocol (Online)

To motivate that the parallelization of circuit garbling provides advantages in Yao's full protocol with pipelining and assuming a fast network connection, we study the protocol for all applications running on two separate cloud instances in the same Amazon region (LAN setting). We observed an average round trip time of 0.6±0.3 ms and a transfer bandwidth of 5.0±0.4 Gbps using `iperf`. Following the results of the offline experiments, we benchmark the more promising CGP approach in the online setting.

To measure the benefits of parallelization, we first benchmark the single core performance of Yao's protocol in the described network environment. Table 5.3 shows the sequential runtime for all applications using two security levels $\kappa = 80$ bit (short term) and $\kappa = 128$ bit (long term). This runtime includes the time spent on the input as well as the output phase. Furthermore, the observed throughput, measured in non-linear gates per second, as well as the required bandwidth are presented. We observe that for security levels of $\kappa = 80$ and $\kappa = 128$ a similar gate throughput is achieved. Consequently, we deduce that in this setup the available bandwidth is not stalling the computation. We also observe that that the time spent on OTs in all applications is practically negligible ($< 5\%$) in comparison to the time spent on circuit garbling.

In Fig. 5.8 the performance gain of CGP is presented. The speed-up is measured in relation to the sequential total runtime. The timing results show that CGP scales almost linearly up to four threads when using $\kappa = 80$ bit labels. Using $\kappa = 128$ bit labels, no further speed-up beyond three threads is noticeable. Thus, the impact of the network limits is immediately visible. Five ($\kappa = 80$ bit), respectively three ($\kappa = 128$ bit) threads are sufficient to saturate the available bandwidth in this experiment. Achieving further speed-ups is impossible without increasing the available bandwidth.

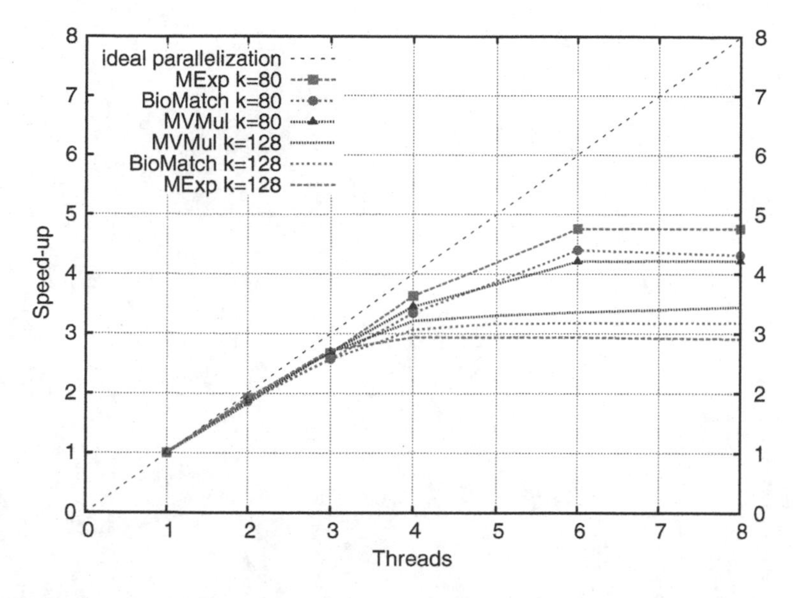

Fig. 5.8 Yao's protocol—CGP performance. Shown is the speed-up of all three applications in the LAN setting with $\kappa = 128$ bit and $\kappa = 80$ bit security

In conclusion, MPC based on Yao's garbled circuits protocol can greatly benefit from compiler assisted parallelization. The FGP approach can be efficient for some circuits, yet its scalability highly depends on the circuit's width. This problem is addressed in more detail in the next chapter, where we study how to compile circuits with a shallow depth. The CGP approach shows a more efficient parallelization, given suitable, i.e., parallel decomposable, applications.

Chapter 6
Compiling Depth-Optimized Circuits for Multi-Round MPC Protocols

6.1 Motivation and Overview

In the previous chapters, we focussed on describing techniques to compile circuits with a minimal number of non-linear gates. However, we also observed that circuits with a shallow depth are of interest for efficient parallelization (Chap. 5). Moreover, one of the first MPC protocols, namely the GMW protocol by Goldreich, Micali and Widgerson [37] and many subsequent protocols, e.g., BGW by Ben-Or et al. [6], Sharemind by Bogdanov et al. [12], SPDZ by Damgard et al. [26], TinyOT by Nielsen et al. [66], or the protocol by Furukuwa et al. [34] have a round complexity that is linear in the circuit depth. Hence, for this class of MPC protocols it is crucial to also consider the circuit depth as a major optimization goal, because every layer in the circuit increases the protocol's runtime by the round trip time (RTT) between the computing parties. This is of special importance, as latency is the only computational resource that has reached its physical boundary. For computational power and bandwidth, parallel resources can always be added. Thus, for these MPC protocols it is much more vital to minimize the depth of circuits, rather than speeding-up the computational efficiency, especially in the asymptotic case.

We support these thoughts with practical numbers. For example, the MPC protocol by Furukuwa et al. [34], which provides security against malicious adversaries assuming an honest majority in an three-party setting, can compute more than one billion gates on a server CPU with 20 cores. At the same time, the network latency between Asia and Europe[1] is in the range of a hundred milliseconds. In such a setting with a latency of 100 ms, a gate-level parallelism of at least 100 Million

[1]Even though, MPC is often benchmarked in a LAN setting, a WAN setting is the more natural deployment model of MPC.

N. Büscher, S. Katzenbeisser, *Compilation for Secure Multi-party Computation*, SpringerBriefs in Computer Science, https://doi.org/10.1007/978-3-319-67522-0_6

Table 6.1 Depth of building blocks. Comparison of the depth of the here presented building blocks with the previously known best constructions

Operation	Previous work [71]	This work
n-bit addition ($n \rightarrow n+1$)	$2\log_2(n)+1$	$\log_2(n)+1$
n-bit multiplication ($n \rightarrow 2n$)	$3\log_2(n)+4$	$2\log_2(n)+3$
m:1 multiplexer	$\log_2(m)$	$\lceil\log_2(\lceil\log_2(m+1)\rceil)\rceil$

gates is required, to avoid stalling the CPU. Therefore, it is worthwhile to study optimization and compilation techniques for the automatic creation of Boolean circuits with minimal depth (or maximal width).

Compiler Extension In this chapter, we describe how CBMC-GC's compilation chain can be adapted to compile circuits with a minimal depth. We implement the techniques as part of a compiler extension to CBMC-GC named *ShallowCC* using a three-folded approach:

First, we describe minimization techniques that operate on the source-code level. For example, we discuss how aggregations, e.g., sum or minima computation over an array, described in a sequential manner can be regrouped in a tree structure. This allows to achieve a circuit with a logarithmic instead of a linear depth. We also portray a technique that detects consecutive arithmetic operations and replaces them by a more efficient dedicated circuit (known as *Carry-Save-Networks*) rather than a composition of multiple individual arithmetic building blocks. Second, we present depth- and size-optimized constructions of major building blocks, e.g., adder and multiplexer, required for the synthesis of larger circuits. An overview of these and comparison with previous constructions is given in Table 6.1. Third, we adapt CBMC-GC's gate-level optimization methods, described in Sect. 4.4, to also consider depth as an important optimization goal.

Chapter Outline In Sect. 6.2 we describe the adapted compile chain for depth-minimization, including a description of building blocks as well as an adapted fix-point minimization algorithm. A detailed experimental evaluation is given in Sect. 6.3.

6.2 Compilation Chain for Low-Depth Circuits

The size-minimized circuits generated by CBMC-GC are not necessarily depth-minimized, as both optimization goals can be orthogonal. Therefore, we introduce multiple techniques that deviate from CBMC-GC's compilation chain to achieve depth-minimized circuits. An implementation of these techniques is realized as

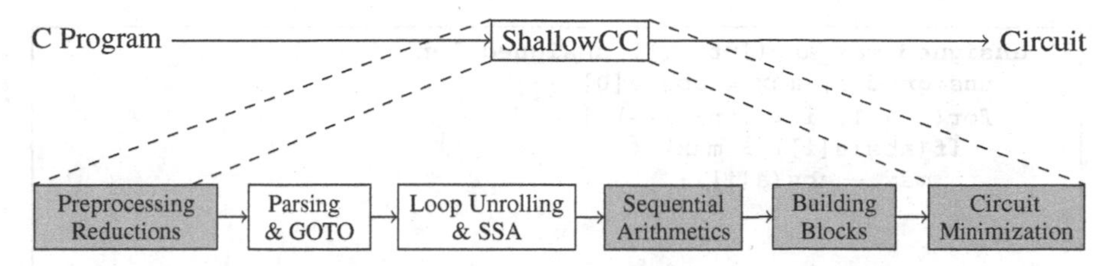

Fig. 6.1 Compilation chain. Shown is ShallowCC's compilation chain from ANSI-C to Boolean circuits. Marked in gray are all modifications to CBMC-GC original compilation chain

a compiler extension to CBMC-GC named ShallowCC. The difference to the compilation chain of CBMC-GC are illustrated in Fig. 6.1 and described in this section.

The extended compilation chain begins with a code preprocessing to detect and transform reduction statements on the source code level (see Sect. 6.2.1). In the next steps the original tool chain of CBMC-GC is applied, the code is parsed and translated into a GOTO program, all bounded loops and recursions are unrolled using symbolic execution and the resulting code is transformed into Single Static Assignment (SSA) form. In the fourth step, the SSA form is used to detect and annotate successive arithmetic statements (see Sect. 6.2.2). Afterwards, all statements are instantiated with depth-minimized building blocks (see Sect. 6.2.3), before a final gate-level minimization takes place (see Sect. 6.2.4).

6.2.1 Preprocessing Reductions

We refer to a reduction as the aggregation of multiple programming variables into a single result variable, e.g., the sum of an array. An exemplary reduction is illustrated in the code example in Listing 6.1. This code computes the maximum norm of a vector. It iterates over an integer array, computes the absolute value of every element and then reduces all elements to a single value, namely their maximum. A straight forward translation of the maximum computation leads to a circuit consisting of `len` − 1 sequentially aligned comparators and multiplexers, as illustrated for four values in Fig. 6.2a. However, the same functionality can be implemented with logarithmic depth when using a tree structure, as illustrated in Fig. 6.2b. Thus, when compiling circuits with minimal depth, it is worthwhile to rewrite sequential reductions. To relieve the programmer from this task, ShallowCC aims at automatically replacing sequential reductions found in loop statements by tree-based reductions.

```
1  unsigned max_abs(int a[], unsigned len) {
2    unsigned i, max = abs(a[0]);
3    for(i = 1; i < len; i++) {
4      if(abs(a[i]) > max) {
5        max = abs(a[i]);
6      }
7    }
8    return max;
9  }
```

Listing 6.1 Maximum vector norm. Exemplary function that computes the maximum norm

```
1  DT reduction_tree(DT *a, unsigned len, (*cmp)(DT, DT)) {
2    unsigned i, step = 1;
3    while(step < len) {
4      for(i = 0; i + step < len; i += (step << 1)) {
5        a[i] = (*cmp)(a[i], a[i + step]) ? a[i] : a[i + step];
6      }
7      step <<= 1;
8    }
9    return a[0];
10 }
```

Listing 6.2 Reduction template. Simplified tree reduction template for an arbitrary array a of length len, as applied when rewriting reductions in loop statements for an arbitrary datatype DT

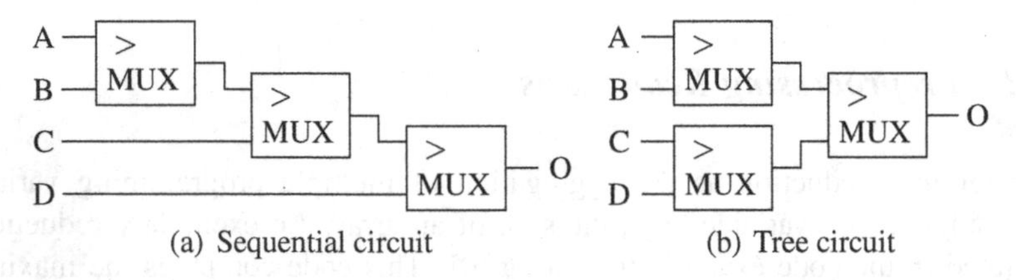

Fig. 6.2 Maximum circuit. Shown are circuit variants to compute the maximum of four values consisting of comparators and multiplexers. The circuit in (**a**) uses a sequential organization of building blocks, whereas the circuit in (**b**) follows a tree structure

Since detecting reductions in loop statements is a common task in automatized parallelization, we can use the compilation techniques presented in Chap. 5. Parallelization frameworks are not only very suited to detect parallelism but also to detect sequential reductions, as these can significantly degrade the performance of parallelization. Therefore, we extend the techniques presented in Chap. 5 to parse reduction annotations and to rewrite the code with *clang* (source-to-source compilation). For this, we first identify the loop range and reduced variable to instantiate a code template that computes the reduction in a tree structure. Such a template is illustrated in Listing 6.2 for an arbitrary comparator function cmp() and datatype DT. A similar template can be used for sum or product computations.

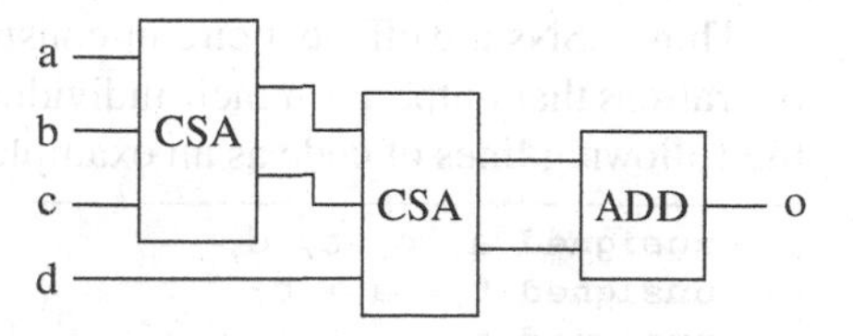

Fig. 6.3 Carry-save network. Addition of four numbers using a CSN consisting of two CSAs and one adder (ADD)

This optimization improves the depth of reductions over m elements from $O(m)$ to $O(\log m)$. To illustrate the effect of this optimization in practice we compile an exemplary circuit that computes the minimum of a 32-bit integer array with 100 elements. In this example, the depth of the compiled circuit reduces from 592 to 42 non-linear gates. Further results are given in Sect. 6.3.3.

6.2.2 *Sequential Arithmetics and Carry-Save Networks (CSNs)*

In the early 1960s [30], Carry-Save Addition was introduced, which allows the fast addition of three or more numbers. The main component of a Carry-Save Addition is the 3:2 Carry-Save Adder (CSA). For three given numbers a, b and c, a CSA computes two numbers x and y, whose sum is equal to the sum of a, b and c, i.e., $a+b+c = x+y$. The key insight is that this computation can be done with significantly less (constant) depth than when computing the real sum of two numbers, which requires a circuit that has a depth logarithmic in the numbers' bit-widths. Given a 3:2 CSA, the sum of k numbers can be computed by first reducing the k numbers to two numbers using a network of $k-2$ CSAs, called Carry-Save Network (CSN) and then computing the final addition using a standard adder. For example, the addition of four numbers a, b, c, d in such a tree is illustrated in Fig. 6.3.

A CSA for three n bit numbers a, b, and c can be constructed by using n parallel Full-Adders (FA, see Sect. 4.3), where the sum bits and carry out bits each form the two numbers x and y. Hence for every bit position $i \in [0, n-1]$, X_i and Y_{i+1} are set to

$$X_i = A_i \oplus B_i \oplus C_i$$
$$Y_{i+1} = (A_i \oplus C_i)(B_i \oplus C_i) \oplus C_i,$$

with $Y_0 = 0$. The sum bit X_i can actually be computed for free and the carry-out bit Y_{i+1} only requires a single non-linear gate. Thus, a CSA has depth and size $s^{nX} = d^{nX} = 1$ in MPC. Using a tree based aggregation and exploiting the property that one output of a CSA can be computed with zero non-linear gates, the partial sums x and y for k numbers can be computed with the depth of $d^{nX}{}_{CSA}(k) = \lceil \log_2(k) - 1 \rceil$ [71].

Thus, CSNs are efficient circuit constructions for multiple successive arithmetic operations that outperform their individual composition in size and depth. Consider the following lines of code as an example:

```
unsigned a, b, c, d;
unsigned t = a + b;
unsigned sum = t + c + d;
```

A straight forward compilation, as in CBMC-GC, leads to a circuit consisting of three binary adders: `sum = ADD(ADD(ADD(a,b), c), d)`. However, if it is possible to identify that a sum of four independent operands is computed, a CSN can be initiated instead: `sum = ADD(CSA(CSA(a,b,c),d))`. In this short example, the circuit's depth reduces from 18 to 7 non-linear gates. Hence, an automatic detection and translation of sequential arithmetic operations into CSNs is highly desirable. Detecting these operations on the gate level is feasible, for example with the help of pattern matching, yet impractically costly considering that circuits reach sizes in the range of billions of gates. Therefore, ShallowCC aims at detecting these successive statements before their translation to the gate level. We do this by utilizing the SSA form, where each variable is written only once. The SSA form allows efficient data flow analyses and as such, also the search for successive arithmetic operations.

We propose a greedy detection algorithm that consists of two parts. First, a breadth-first search from output to input variables is initiated. Whenever an arithmetic assignment is found, a second backtracking algorithm is initiated to identify all preceding (possibly nested) arithmetic operations. This second algorithm stops whenever a guarded or non-arithmetic statement is found. Once all preceding inputs are identified, the initial assignment can be replaced by a CSN. After every replacement, the search algorithm continues its search towards the input variables. We note that this greedy replacement approach is depth-minimizing, yet not necessarily size optimal. This is because intermediate results in nested statements may be computed multiple times. A trade-off between size and depth is possible by only instantiating CSNs for non-nested arithmetic statements.

Quantifying the improvements in depth reduction and assuming that the addition of two numbers requires a circuit of depth $d^{nX}{}_{Add}$, we observe that by sequential composition $m > 2$ numbers can be added with depth $(m-1) \cdot d^{nX}{}_{Add}$. When using a tree-based structure the same sum can be computed with a depth of $\lceil \log_2(m) \rceil \cdot d^{nX}{}_{Add}$. However, when using a CSA, m numbers can be added with a depth of only $\lceil \log_2(m) - 1 \rceil + d^{nX}{}_{Add}$.

We remark that CSNs can not only be constructed for additions but also for any mix of subsequent multiplications, additions and subtractions, because every multiplication internally consists of additions of partial products and a subtraction is an addition with the bitwise inverse. To illustrate the improvement in practice, for the exemplary computation of a 5×5 matrix multiplication, an improvement in depth of more than 60% can be observed when using a CSN, see Sect. 6.3.3.

6.2.3 Optimized Building Blocks

As mentioned in Sect. 4.3, optimized building blocks are an essential part when designing complex circuits. They facilitate efficient compilation, as they can be highly optimized once and subsequently instantiated at practically no cost during compilation. In the following paragraphs, we present building blocks optimized for depth (and size) for basic arithmetic and control flow operations.

Adder An n-bit *adder* takes two bit strings a and b of length n, representing two (signed) integers, as input and returns their sum as an output bit string s of length $n+1$. The standard adder, described in Sect. 4.3, is the Ripple Carry Adder (RCA) that consists of a successive composition of n FAs. It has a circuit size and depth $s^{nX}{}_{RCA} = d^{nX}{}_{RCA} = O(n)$ that is linear in its bit-width. For faster addition, parallel Prefix Adders (PPAs) are widely used in logic synthesis. Their core concept is to use a tree based prefix network to achieve a logarithmic depth under a size trade-off. Various different tree structures have been proposed. We first discuss their general design before comparing three different tree structures.

In general, the PPA design distinguishes two signals named the generate $G_{i:j}$ and the propagate $P_{i:j}$ signal, used to handle the fast propagation of carry-out bits. A generate signal announces whether the prefix sum of the two input strings $A_iA_{i-1}\ldots A_j + B_iB_{i-1}\ldots B_j$ will generate a carry-out bit. A propagate signal over the same range announces that a carry-out bit will be propagated if additionally a carry-in bit is set. A complete PPA consists of three parts. First, the initial generate and propagate signals are computed over pairs of input bits as $G_{i:i} = A_i{\cdot}B_i$ (carry out bit) and $P_{i:i} = A_i \oplus B_i$ (sum bit). Second, subsequent generate and propagate signals are typically computed in hardware synthesis using the following formula [39]:

$$(G_{i:j},\ P_{i:j}) = (G_{i:k} + P_{i:k}G_{k-1:j},\ P_{i:k}P_{k-1:j}).$$

Finally, in the last step each output bit S_i can then be computed as the XOR of the generate and propagate bit $S_i = P_{i:-1} \oplus G_{i-1:-1}$. We observe that the initial and output phase have a constant depth and that the depth of the intermediate phase depends on the structure of the used parallel prefix network. A close look at the generate and propagate signals reveals, that only one of the two can be set for every bit range. Hence, in circuit design for MPC it is very reasonable to replace the OR(+) by an XOR($\oplus$) in the formula of the generate signal:

$$(G'_{i:j},\ P_{i:j}) = (G'_{i:k} \oplus P_{i:k}G'_{k-1:j},\ P_{i:k}P_{k-1:j}).$$

This insight *halves* the circuit depth of every generate signal, as only one instead of two non-linear gates are required. Consequently, this also halves the depth of the parallel prefix network. A proof of the correctness of this substitution is given in [17].

Applying this idea to the Sklansky PPA design, which is the fastest PPA structure according to the taxonomy by Harris [39], we achieve a construction with a depth

Table 6.2 Adders. Comparison of circuit size s^{nX} and depth d^{nX} of the standard RCA, the previously best known depth-minimized Ladner-Fischer PPA [71], and the Brent-Kung and Sklansky PPAs with the described generate signal optimization

	Depth d^{nX}				Size s^{nX}			
Bit-width	n	16	32	64	n	16	32	64
Ripple-Carry	$n-1$	15	31	63	$n-1$	15	31	63
Ladner-Fischer[28, 71]	$2\lceil\log(n)\rceil+1$	9	11	13	$1.25n\lceil\log(n)\rceil+2n$	113	241	577
Brent-Kung-opt	$2\lceil\log(n)\rceil-1$	7	9	11	$3n$	48	96	192
Sklansky-opt	$\lceil\log(n)\rceil+1$	5	6	7	$n\lceil\log(n)\rceil$	64	160	384

of $d^{nX}{}_{Sk}(n) = \lceil\log_2(n)\rceil + 1$ and a size of $s^{nX}{}_{Sk} = n\lceil\log_2(n)\rceil$ for an input bit length of n and output bit length of $n+1$.

In Table 6.2 a depth and size comparison with the standard Ripple-Carry adder, the Ladner-Fischer adder described by Schneider and Zohner [71], the Sklansky PPA, and an alternative PPA structure, namely the Brent-Kung PPA is given for different bit-widths. We observe that the RCA provides the smallest size and the Sklansky adder the shallowest depth. The Brent-Kung adder provides a trade-off between size and depth. The here optimized Sklansky and Brent-Kung adder significantly outperform the previous best known depth-minimized construction in size and depth.

Subtractor As described in Sect. 4.3, a subtractor can be implemented using an adder and one additional non-linear gate with the help of the two's complement representation, which is $-a = \overline{a}+1$ with $\overline{a}$ being the inverted binary representation. Consequently, $a-b$ can be represented as $a+\overline{b}+1$. Hence, the subtractor profits to the same degree from the optimized addition.

Comparator A depth-minimized *equivalence (EQ)* comparator can be implemented by a tree based OR composition over pairwise XOR gates to compare single bits, similar to the size-minimal construction presented in Sect. 4.3. This yields a depth of $d^{nX}{}_{EQ}(n) = \lceil\log_2(n)\rceil$ and size of $s^{nX}{}_{EQ}(n) = n-1$ gates [71].

A depth-minimized *greater-than (GT)* comparator can be implemented in the same way, as the size-minimized GT (see Sect. 4.3) comparator by using the observation that $x > y \Leftrightarrow x - y - 1 \geq 0$ and returning the carry out bit. This approach leads to a circuit depth that is equivalent to the depth of a subtractor.

Multiplier In the size-minimized multiplication (see Sect. 4.3), n partial products of length n are computed and then added. This approach leads to a quadratic size $s^{nX}{}_{MUL,s} = 2n^2-n$ and linear depth $d^{nX}{}_{MUL,s} = 2n-1$. A faster addition of products can be achieved when using CSAs. Such a tree based multiplier consists of three steps: First, the computation of all $n \times n$ partial products, then their aggregation in a tree structure using CSAs, before the final sum is computed using a two-input adder. The first step is computed with a constant depth of $d^{nX}{}_{PP} = 1$, as only one single AND gate is required. For the last step, two bit strings of length $2n-1$ have to be added. Using the optimized Sklanksy adder, this addition can be realized in

Table 6.3 Multipliers. Comparison of circuit depth d and size s of the school method, the multiplier given in [71] and our optimized Wallace construction

	Depth d^{nX}				Size s^{nX}			
Bit-width	n	16	32	64	n	16	32	64
Standard	$2n-1$	45	93	189	n^2-n	496	2016	8128
MulCSA [71]	$3\lceil\log_2(n)\rceil+4$	16	19	22	$\approx 2n^2+1.25n\log_2(n)$	578	2218	8610
Wallace-opt	$2\lceil\log_2(n)\rceil+3$	11	13	15	$\approx 2n^2+n\log_2(n)$	512	2058	8226

${d^{nX}}_{Sk}(n) = \lceil\log_2(2n-1)\rceil+1$. The second phase allows many different designs, as the CSAs can arbitrarily be composed. The fastest composition is the Wallace tree [74], which leads to a depth of ${d^{nX}}_{CSA}(n) = \log_2(n)$ for MPC. Combing all three steps, a multiplication can be realized with depth

$${d^{nX}}_{Wa}(n) = {d^{nX}}_{PP} + d_{CSA}(n) + {d^{nX}}_{Sk}(2n-1) = 2\lceil\log_2(n)\rceil + 3.$$

In Table 6.3 we present a comparison of the multipliers discussed above with the depth optimized one presented in [71]. Compared with this implementation, we are able reduce the depth by at least a third for any bit-width.

Multiplexer We recap Sect. 4.3, where we described a 2:1 MUX that only requires one single non-linear gate for every pair of input bits by computing the output as $O = (D^0 \oplus D^1)C \oplus D^0$. Hence, a 2:1 n-bit MUX has size ${s^{nX}}_{MUX}(n) = n$ and depth of ${d^{nX}}_{MUX}(n) = 1$. A 2:1 MUX can be extended to a m:1 MUX that selects between m input strings $D^0, D^1, \ldots, D^m$ using $\log_2(m)$ control bits $C = C_0, C_1, \ldots C_{\log(m)}$ by tree based composition described in detail Sect. 4.3. This circuit construction has a size of ${s^{nX}}_{MUX\_tree}(m,n) = (m-1)\cdot {s^{nX}}_{MUX}(n)$ and a depth that is logarithmic in the number of data inputs ${d^{nX}}_{MUX\_tree}(m,n) = \log_2(m)$.

A further depth-minimized m:1 MUX can be constructed under a moderate size trade-off, by using a design that is similar to a *disjunctive normal form (DNF)* over all combinations of choice bits. Every conjunction of the DNF encodes a single choice together with the associated data wire. For MPC, this construction leads to a very low depth, because the disjunctive ORs can be replaced by XORs, as all choices are mutually exclusive. For example, a 4:1 MUX is then expressed as

$$O = D^0\overline{C_0}\overline{C_1} \oplus D^1\overline{C_0}C_1 \oplus D^2C_0\overline{C_1} \oplus D^3C_0C_1.$$

Consequently the depth of a depth-minimized DNF m:1 MUX (MUX_{DNFd}) is equivalent to the depth of a single conjunction. A single conjunction can be computed in a tree-based manner. For example one conjunction in a 128:1 MUX can be encoded as $(((D_{127}C_0)(C_1C_2))((C_3C_4)(C_5C_6)))$, which leads to depth that is logarithmic in the number of control bits:

$${d^{nX}}_{MUX\_DNFd}(m,n) = \lceil\log_2(\lceil\log_2(m)\rceil+1)\rceil.$$

Unfortunately, a straightforward implementation of m:1 MUX_{DNFd} over data inputs with a width of n-bit that follows the description above has a size that is $\log(m)$ times larger than the size of a tree based MUX:

$$s^{nX}{}_{MUX\_DNFd}(m,n) = \log(m) \cdot mn.$$

This increase by a logarithmic factor can be quite significant for larger m. Therefore, we describe a second DNF based construction, referred to as MUX_{DNFs}, which offers a more practical size-depth trade-off. The idea is to first compute every conjunction of choice bits in a tree-based manner, before AND-gating each "encoded" choice with the data input. Considering the same 127:1 MUX example, a conjunction is encoded as $D_{127}((C_0(C_1C_2))((C_3C_4)(C_5C_6)))$. This construction has a very similar depth than the previously described:

$$d^{nX}{}_{MUX\_DNFs}(m,n) = \lceil \log_2(\lceil \log_2(m) \rceil) \rceil + 1.$$

However, with the separation of control and data inputs, the computation of various combinations of choice bits can be merged more effectively between different conjunctions, e.g., the choices $C_0C_1C_2$ and $\overline{C_0}C_1C_2$ require both the computation of C_1C_2. This reduces the size to:

$$\begin{aligned} s^{nX}{}_{MUX\_DNFs}(m,n) &= \text{control combinations} + \text{AND gating with data inputs} \\ &= \left(m + \frac{m}{2^1} + \frac{m}{2^2} + \cdots + \frac{m}{2^{m-2}}\right) + mn \\ &< 2m + mn. \end{aligned}$$

In Table 6.4 a comparison of the three MUXs is given for a different number of inputs m and a typical bit-width of 32 bits. In summary, we improved the depth of MUXs to $O(\log(\log(m)))$, which is almost constant in practice, with a moderate increase in size.

Table 6.4 Multiplexers. Exemplary comparison of circuit depth d and size s of m:1 multiplexers for a different number of inputs m of bit-width $n = 32$

	Depth d^{nX}				Size s^{nX}			
Input choices	m	8	128	1024	m	8	128	1024
MUX_{Tree}	$\lceil \log(m) \rceil$	3	7	10	$(m-1) \cdot n$	244	4064	31,968
MUX_{DNFd}	$\lceil \log_2(\lceil \log_2(m) + 1 \rceil) \rceil$	2	3	4	$mn \cdot \lceil \log(m) \rceil$	768	28,672	320,000
MUX_{DNFs}	$\lceil \log_2(\lceil \log_2(m) \rceil) \rceil + 1$	3	4	5	$2m + mn$	272	4352	34,000

6.2.4 Gate Level Minimization Techniques

Minimizing the circuit on the gate level is last step in CBMC-GC's compilation chain. For ShallowCC, we left parts, e.g., structural hashing and SAT sweeping, of the fix-point minimization algorithm unmodified, because they do not have a dedicated depth-minimizing counterpart and both already contribute in reducing the circuit complexity and thus, also contribute to reducing the circuit depth. Instead, we adapt the template based rewriting phase.

Circuit rewriting in CBMC-GC only considers patterns that are size decreasing and have a depth of at most two binary gates. For depth reduction, however, it is useful to also consider deeper circuit structures, as well as patterns that are size preserving but depth decreasing. For example, the sequential structure $X = A + (B + (C + (D + E)))$ can be replaced by the tree based structure $X = ((A + B) + C) + (D + E)$ with no change in circuit size. Therefore, in ShallowCC we extend the rewriting phase by several depth-minimizing patterns, which are not necessarily size decreasing. In total 21 patterns changed, resulting in more than 80 patterns that are searched for. Furthermore, circuit rewriting as described in Sect. 4.4 applies a pattern matching algorithm, which searches for pattern by pattern, always matching them as a whole. However, to replace sequential structures by tree based structures, it is worthwhile to consider arbitrary sequential compositions of gates. Therefore, we replaced the fixed pattern matching algorithm by a flexible and recursive search for sequential structures. For example, instead of having dedicated patterns for $X = A \cdot (B \cdot (C \cdot (D \cdot E)))$ and $X = A \cdot (B \cdot C \cdot (D \cdot E))$, both are matched with the recursive algorithm that replaces all possible sequential structures of non-linear gates that are free of intermediate outputs.

Finally, we modify the termination condition of the fix-point optimization routine such that the algorithm only terminates if no further size *and depth* improvements are made (or a user defined time limit is reached). Moreover, for performance reasons, the rewriting first only applies fixed depth patterns, before applying the recursive search for deeper sequential structures.

Quantifying the improvements of individual patterns is almost impossible. This is because the heuristic approach commonly allows multiple patterns to be applied at the same time and every replacement has an influence on future applicability of further patterns. Nevertheless, the whole set of patterns that we identified is very effective, as circuits before and after gate level minimization differ up to a factor of 20 in depth, see Sect. 6.3.3.

6.3 Experimental Evaluation

We evaluate the effectiveness of automatized depth-minimization in three directions. First, we compare the circuits generated by ShallowCC with circuits that have been optimized for depth by hand or using state-of-the-art hardware synthesis tools.

Then, we study the effectiveness of different implemented optimization techniques individually. Finally, we show that depth-minimized circuits, even under size trade-offs, significantly reduce the online runtime of multi-round MPC protocols for different network configurations. We begin by describing the parameters of the benchmarked functionalities.

6.3.1 Benchmarked Functionalities and Their Parameters

For comparison purposes, we focus on functionalities that have been used before to benchmark MPC and that have also partly been used in previous chapters (details can be found in Sect. 2.3). The applications used to evaluate ShallowCC are:

- *Integer arithmetics.* Due to their importance in almost every computational problem, we benchmark arithmetic building blocks individually. For multiplication we distinguish results for output bit strings of length n and of length $2n$ (overflow free) for n-bit input strings.
- *Floating point arithmetics.* We abstain from implementing hand-optimizing floating point circuits, but instead rely on CBMC-GC's capabilities to compile a IEEE-754 compliment software floating point implementation of addition (*FloatAdd*) and multiplication (*FloatMul*) written in C.
- *Distances.* We study the *Hamming* distance over 160 and 1600 bits. Moreover, we study the *Manhattan* distance for two integers of bit-width 16 and 32 bit, and we also study the squared two dimensional *Euclidean* distance for the same bit-widths.
- *Matrix multiplication (MMul).* MMul is a purely arithmetic task, and therefore a good showcase to illustrate the automatic translation of arithmetic operations into CSNs to achieve very low depth. Here we use a matrix of size 5×5 and 32 bit integers.
- *Oblivious arrays.* Oblivious data structures are a major building block for the implementation of privacy preserving algorithms. The most general data structure is the oblivious array that hides the accessed index. We benchmark the read access to an array consisting of 32 integers of size 8 bit and 1024 integers of size 32 bit.
- *Biometric matching (BioMatch).* As in previous chapters, we use the squared Euclidean distance as distance function between two samples. Moreover, we use a database of size of $n = 32$ and $n = 1024$, and $d = 4$ features per sample with an integer bit-width of $b = 16$ using overflow free multiplication.

6.3.2 Compiler Comparison

We compiled all applications with ShallowCC on an Intel Xeon E5-2620-v2 CPU with a minimization time limit of 10 min.

The resulting circuit dimensions, when compiling the benchmark functionalities, for different parameters and bit-widths are shown in Table 6.5. Furthermore, the circuit size, when compiled with the size-minimizing Frigate compiler [61] and CBMC-GC v0.9 is given, as well as a comparison with the depth-minimized circuit constructions of [28] and [71]. The results for [28, 61] and [71] are taken from the publications.

Comparing the depth of the circuits compiled by ShallowCC with the hand and tool minimized circuits of [28, 71] we observe a depth reduction at least 30% for most functionalities. The only exception are the floating point operations, which do

Table 6.5 Compiler comparison. Comparison of circuit size s^{nX} and depth d^{nX} when compiling functionalities with the size-minimizing Frigate [61] and CBMC-GC v0.9 compiler, the best manually depth-minimized circuits given in [28, 71] and the circuits compiled by ShallowCC. Improvements are computed in comparison with the best previous work [28, 71]. The '–' indicates that no results were given. Marked in bold face are cases with significant depth reductions

		Size-minimized			Depth-minimized				
		Frigate	CBMC-GC		Prev. [28, 71]		ShallowCC		Improv.
Circuit	n	s^{nX}	s^{nX}	d^{nX}	s^{nX}	d^{nX}	s^{nX}	d^{nX}	d^{nX}
Building blocks									
Add $n \rightarrow n$	32	31	31	31	232	11	159	**5**	**54%**
Sub $n \rightarrow n$	32	31	61	31	232	11	159	**5**	**54%**
Mul $n \rightarrow 2n$	32	2082	4600	67	2218	19	2520	**15**	**21%**
Mul $n \rightarrow n$	64	4035	4782	67	–	–	4350	**16**	–
Arithmetics									
Div	32	1437	2787	1087	7079	207	5030	**192**	**7%**
MMul 5x5	32	128,252	127,225	42	–	–	128,225	**17**	–
FloatAdd	32	–	2289	164	1820	59	2437	62	−5%
FloatMul	32	–	3499	134	3016	47	3833	54	−14%
Distances									
Hamming-160	1	719	371	9	–	–	281	**7**	–
Hamming-1600	1	4691	7521	31	–	–	1021	**12**	–
2D-Euclidean	16	–	826	47	1171	29	1343	**19**	**34%**
2D-Euclidean	32	–	3210	95	3605	34	5244	**23**	**32%**
2D-Manhattan	16	–	187	31	296	19	275	**13**	**31%**
2D-Manhattan	32	–	395	63	741	23	689	**16**	**30%**
Privacy preserving protocols									
BioMatch-32	16	–	88,385	1101	–	–	90,616	**55**	–
BioMatch-1024	16	–	2.9M	35,821	–	–	2.9M	**90**	–
Ob.Array-32	8	–	803	66	248	5	538	**3**	**40%**
Ob.Array-1024	32	–	100,251	2055	32,736	10	65,844	**4**	**60%**

not reach the same depth as given in [28]. This is because floating point operations mostly consist of bit operations, which can significantly be hand optimized on a gate level, but are hard to optimize when complied from a high-level implementation in C. When comparing circuit sizes, we observe that ShallowCC is compiling circuits that are competitive in size to the circuits compiled by size minimizing compilers. A negative exception is the addition, which shows a significant trade off between depth and size. However, the instantiation of CSNs allows ShallowCC to compensate these trade-offs in applications with multiple additions, e.g., the matrix multiplication. In Sect. 6.3.4 we analyze these trade-offs in more detail. In summary, ShallowCC is compiling ANSI-C code to Boolean circuits that outperform hand crafted circuits and tool optimized circuits in depth, with moderate increases in size. This also illustrates that the here proposed combination of minimization techniques outperforms the classic hardware synthesis tool chain used in [28].

6.3.3 *Evaluation of the Optimizations Techniques*

Table 6.6 shows the result of an evaluation of the different optimization techniques for various example functionalities. For every functionality the same source code is compiled twice, once with the specified optimization technique enabled and once without. Obviously, not all optimizations apply to all functionalities, therefore, we only investigate a selection of functionalities that profit from the different optimization techniques. The CSN detection shows its strengths for arithmetic functionalities. For example, the 5×5 matrix multiplication shows a depth reduction of 60%, when optimizations are enabled. This is because the computation of a single vector element can be grouped into one CSN. The detection of reductions is a very specific optimization, yet, when applicable, the depth saving can be

Table 6.6 Effectiveness of optimization techniques. Comparison of circuit dimensions when compiled by ShallowCC with different optimization techniques enabled or disabled. Marked in bold face are significant improvements

		W/o optimization		W/ optimization		Improvement	
Circuit	n	Size s^{nX}	Depth d^{nX}	size s^{nX}	Depth d^{nX}	size s^{nX}	Depth d^{nX}
Optimization: carry-save networks CSNs							
MMul 5×5	32	143,850	42	128,225	17	**11%**	**60%**
4D-EuclidianDst	16	2993	40	2459	20	**18%**	**50%**
Optimization: reduction							
Minima-100	16	5742	594	5742	42	0%	**92%**
BioMatch-1024	16	2,9M	7181	2,9M	90	0%	**98%**
Optimization: gate level minimization							
Hamming-160	1	5389	77	281	7	**95%**	**88%**
FloatAdd	32	10,054	194	2431	74	**75%**	**61%**

significant. When computing the minima of 100 integers, a depth reduction of 92% is visible. Note that in this test the circuit size itself is unchanged, as only the order of multiplexers is changed. Gate level minimization is the most important optimization technique for all functionalities, which do not use all bits available in every program variable. In these cases constant propagation applies, which leads to significant reductions in size and depth, as exemplary shown for the floating point addition and computation of the Hamming distance. In general, when applicable, the optimization methods significantly improve the compilation result of ShallowCC.

6.3.4 Protocol Runtime

To show that depth-minimization improves the online time of MPC protocols, we evaluate a selection of circuits in the ABY framework [29]. ABY provides a state-of-the-art two-party implementation of the GMW protocol [37] secure in the semi-honest model (a protocol description is given in Sect. 2.2.3). We extended the ABY framework by an adapter to parse CBMC-GC's circuit format. For our experiments, we connected two machines, which are equipped with an AMD FX 8350 CPU and 16GB of RAM, running Ubuntu 15.10 over a 1Gbit ethernet connection in a LAN. To simulate different network environments we made use of the Linux network emulator *netem*.

In this experiment the *online* protocol runtimes of size and depth minimized circuits for different RTTs are compared. *netem* simulates a RTT by stalling the transmission of packages accordingly. In our results, we omit timings for the pre-processing *setup* phase, as this pre-computation can take place independently of the evaluated circuits and with any degree of parallelism. We ran this experiment for different RTTs, starting with zero delay up to a simulated RTT of 80 ms.

The first functionality that we investigate is the biometric matching application with a database of 1024 entries. Here, we compare the circuits generated by CBMC-GC with and without ShallowCC enabled. The resulting circuit dimensions are given in Table 6.5, and the averaged (ten runs) protocol runtimes are given in Fig. 6.4. We observe speed-ups of the depth-minimized circuit over the size-minimized circuit of a factor between 2 and 400, when increasing the RTT from ~1 to 80 ms.

The second functionality that we evaluate is the array read (MUX), which allows to analyze a size-depth trade-off. We compiled the read access to an array with 1024 integers of size 32 bit. We compare the tree based MUX, as proposed in [71] with depth $d^{nX} = 10$ and size $s^{nX} = 32{,}736$ with our depth optimized MUX_{DNFd}, which has a depth of $d^{nX} = 4$ and size $s^{nX} = 65{,}844$ after gate level minimization. Each circuit is evaluated with ABY individually, as well as 100 times in parallel. This also allows to investigate whether single instruction multiple data (SIMD) parallelism, which is favored in GMW [29], has a significant influence on the results. The resulting online runtimes for both circuits are illustrated in Fig. 6.5. All data points are averaged over 100 runs. We observe that for almost every network configuration beyond 1 ms RTT, the depth optimized circuits outperform their size-

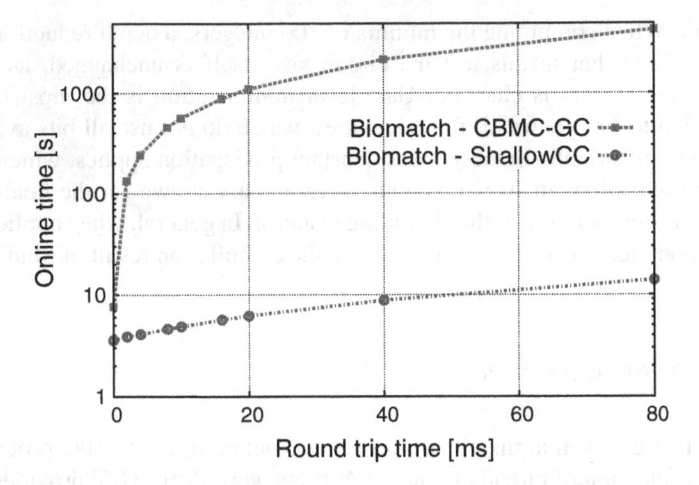

Fig. 6.4 GMW protocol runtime. Shown is the runtime when evaluating a depth (ShallowCC) and size (CBMC-GC) minimized biometric matching application. We observe that the depth optimized circuit significantly outperforms the size optimized circuit for any noticeable RTT

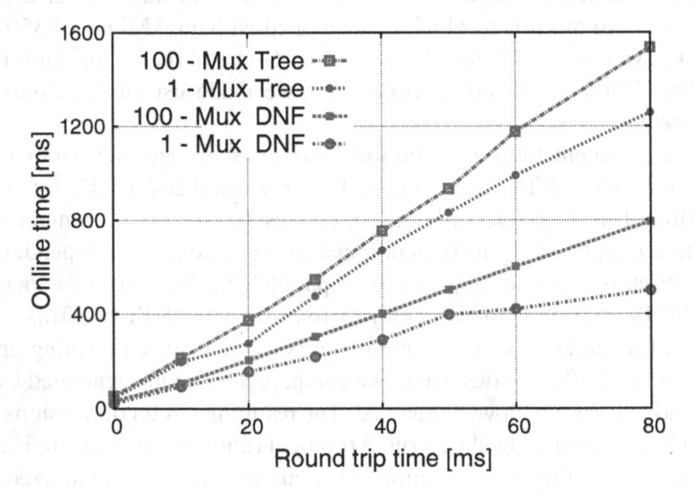

Fig. 6.5 Multiplexer comparison. GMW protocol runtime when evaluating the depth-minimized DNF and size-minimized tree 1024:1 MUX for a single and parallel array read access. The DNF based MUX significantly outperforms the tree based MUX for any noticeable RTT

optimized counterparts by a factor of two. The reason for the factor of two is, that the GMW protocol requires one communication round for input sharing as well as one round for output sharing, which leads to six communication rounds in total for the MUX_{DNFd} and 12 rounds for the MUX_{tree}. Moreover, we observe that the here applied data parallelism shows no significant effect on the speed-up gained through depth reduction.

In conclusion, the experiments support our introductory statement that depth minimization is of uttermost importance to gain further speed-ups in round-based MPC protocols. With the ShallowCC extension to CBMC-GC we are capable to achieve these speed-ups by compiling depth-minimized circuits that are up to 2.5 times shallower than hand optimized circuits and up to 400 times shallower than circuits compiled from size optimizing compilers.

Chapter 7
Towards Scalable and Optimizing Compilation for MPC

7.1 Motivation and Overview

In Chap. 4 we observed that for the creation of efficient MPC applications, optimizing compilers are needed. The approach of CBMC-GC emphasizes optimization. Yet, the drawback of the optimization approach adopted CBMC-GC is its limited scalability. For example, Pullonen and Siim [69] reported a compile time of more than 1 h for a circuit consisting of only 400,000 gates, using an early release of CBMC-GC. This is undesirable, as nowadays MPC implementations can handle multiple millions of gates per second (e.g., [5] for Yao's garbled circuits or [1] for GMW). Consequently, practical applications can scale to billions of gates. Hence, there is a need for scalable compilers that produce reliable and efficient circuit descriptions.

As described in Sect. 1.2, compilers have been proposed that are capable of compiling larger applications than CBMC-GC, e.g., PCF [52] or Frigate [61], by following a strict decomposition paradigm. These compilers decompose the input source into multiple parts that are compiled individually. However, these approaches have the drawback, that they only provide very limited source code and gate level optimization methods. For example, constant arguments in function calls are not propagated, which is essential to compile efficient circuits.

Therefore, in this chapter, we sketch how CBMC-GC can be extended in order to *compile and optimize large programs* in a scalable manner. Thus, we illustrate that scalability through decomposition and circuit optimization are not mutually exclusive. To achieve this goal, we present a multi-pass compilation approach that uses *source code guided optimizations*: First, the input source code is analyzed and decomposed in multiple parts using static analysis (very similar to the parallel decomposition presented in Sect. 5.3.2, yet, here no data parallelism is required). Second, all parts are translated into unoptimized circuits. These smaller *sub-circuits* can later be recombined to resemble the original functionality. Third, based on the translation results, the effect of gate level minimization on every sub-circuit is

N. Büscher, S. Katzenbeisser, *Compilation for Secure Multi-party Computation*, SpringerBriefs in Computer Science, https://doi.org/10.1007/978-3-319-67522-0_7

estimated. These estimates are then used to optimize the sub-circuits with different efforts, i.e. computation time, focussing on hot-spots that are the most promising candidates for optimization. Finally, the optimized circuits are exported using a descriptive file format, which allows their recombination during runtime with any MPC framework.

To illustrate the practicality of our approach, we present a prototype extension to CBMC-GC, which compiles functionalities with millions of gates that are up to 70% smaller than circuits compiled using previously best known scalable compilers. Note that, at the time of writing, this extension is not part of the current open-source release of CBMC-GC.

Chapter Outline We discuss the adapted compilation chain of CBMC-GC, including a description of its implementation in Sect. 7.2. An experimental evaluation is given in Sect. 7.3.

7.2 Adapted Compilation Chain

The size and depth minimization methods of CBMC-GC (see Sect. 4.4) are only applied to the fully unrolled combinational circuit while the methods themselves only work with a local scope. Hence, knowledge about the source code structure is abandoned during the minimization phase, which leads to limited scalability when compiling larger applications. For example, a loop body that is executed k times will be optimized k times because the information about the inherent equivalence of the loop body in each iteration is not transferred to the gate level optimizer. Similarly, functions are compiled and optimized every time they are called. Spending the same time only once, the compilation efficiency and thus the scalability of the compiler can significantly be improved.

In this section, we describe a compilation architecture using static analysis to overcome the limitations of CBMC-GC. Our main idea is to use the information on the source code level to guide the optimization on the gate level more effectively. This approach leads to efficient circuits under reasonable computational costs, as we show in Sect. 7.3. For this purpose, we use a source decomposition approach based on static analysis.

7.2.1 *Compilation Architecture*

The compilation framework consists of the following five components:

Source Code Decomposer The decomposer takes ANSI-C source code with annotations for MPC as input and decomposes (slices) the given code into multiple, independent code files. Slicing criteria are functions and loop bodies. The main functionality and the different parts are interconnected via newly introduced input and output variables that are later translated into wires on the circuit layer. For recombination, a description of the interface between all parts is produced.

Circuit Translator The circuit translator compiles the decomposed code into circuits. Therefore, every code part is translated into an unoptimized circuit by instantiating pre-defined building blocks, as described in Sect. 4.3. In this step, a property description of the generated circuit, including information about the size and the fraction of non-linear gates is exported for the later optimization steps.

Optimization Engine The optimization engine takes the different circuits and their property descriptions as input and assigns optimization budgets (computing time) to the individual circuits. This assignment is based on parameters, such as the circuit size as well as the success of previous optimizations. Hence, the engine coordinates all gate level optimization of the circuits.

Gate Level Optimizer The optimizer takes a circuit description and a optimization budget as input and applies the circuit minimization methods described in Sect. 4.4 to reduce the circuit complexity. The circuit is minimized either until the given budget is reached or no further improvements can be made. If the budget is reached, information about the last minimization pass will be returned to the engine. This information is then used by the optimization engine to assign the remaining optimization budget available for all circuits. Consequently, some sub-circuits will be optimized over multiple passes. More details on the distribution of optimization budget are given in Sect. 7.2.3.

Composer The (re-)composer takes the optimized circuits, as well as the description of their interfaces as input, and merges them into a single circuit. We note, that the composer does not necessarily need to be part of the compile chain. It can also be part of any MPC framework to build the full circuit only during the protocol's execution, which is similar to an integrated compilation approach.

Figure 7.1 illustrates this framework for an exemplary code, that is decomposed into three parts. Each part is compiled into a sub-circuit and optimized individually. In the final step all sub-circuits are merged and composed into a single final circuit.

7.2.2 Global Constant Propagation

The framework described above allows a significantly faster compilation than the unguided, holistic circuit optimization approach. However, a naïve decomposition might have negative consequences on the circuit size, compared to the holistic approach. This is because information about constants are not propagated between the different code parts. Consider the following simplified example source in Listing 7.1, where a function is called with a constant argument:

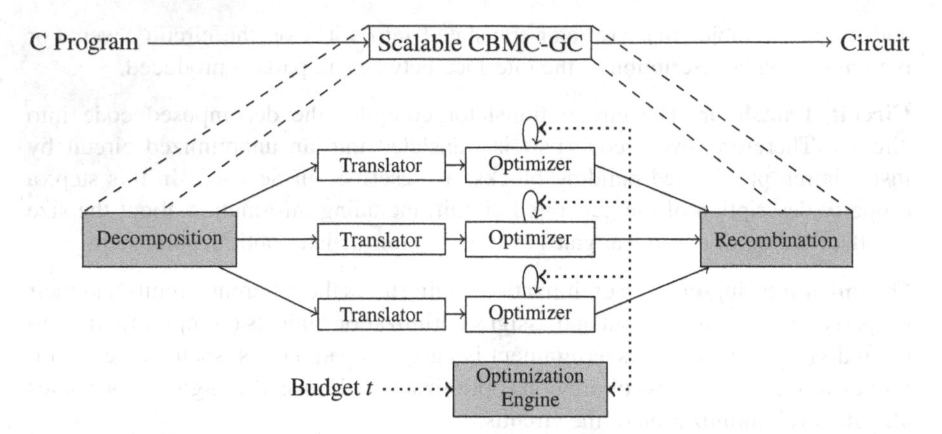

Fig. 7.1 Scalable compilation chain for CBMC-GC. The compilation chain illustrated from source file to circuit. First, the given input code is decomposed, in this example in three parts. Each part is translated into a Boolean circuit. The individual circuits are than optimized with different computational efforts and over multiple iterations, orchestrated by the optimization engine, which assigns the optimization budget depending on various parameters. After optimization, all subcircuits are merged into a single circuit that resembles the input functionality

```
1 unsigned mul(unsigned a, unsigned b) {
2   return a * b;
3 }
4 void main() {
5   /*...*/
6   mul(INPUT_A_x, 5);
7 }
```

Listing 7.1 Contant propagation. Example function call with constant argument

A straightforward decomposition would separate the `mul()` function with two arguments from the `main()` function. The `mul()` function would then be translated into a circuit consisting of a multiplication building block with gate costs that are in $O(n^2)$ with n being the input bit-width. However, without decomposition CBMC-GC would inline the function and compile a multiplication by constant, which requires only 29 instead of 993 non-linear gates for this example over two 32 bit integers. To avoid loosing this efficiency through decomposition, we propose a combination of two strategies.

First, before the decomposition, constant folding and propagation on the source code is applied. Constant folding evaluates constant expressions during compile time and constant propagation uses symbolic execution to push constants as close to the input variables as possible. Second, these constant propagation techniques can be extend on the bit level as follows. During the circuit minimization of every decomposed functionality, the optimization engine is informed about wires with constant outputs and propagates this information between multiple sub-circuits. Thus, whenever an output of a sub-circuit is identified to be constant or equivalent to another output gate of the same sub-circuit, the engine propagates this information to other sub-circuits for the next optimization pass.

7.2.3 Implementation

In this section, we describe an implementation of the described components as compiler extension prototype. Our implementation uses the Frama-C [24] static analysis platform for C, its value analysis plugin [19], which implements powerful abstract interpretation, as well as the PIPS [45] source-to-source parallelization framework.

Decomposer The decomposer first applies the described constant propagation mechanisms on the source code level with the help of Frama-C. Next, inputs and outputs of every function are detected. Given this information, the source code can be sliced for every function call. This is realized by annotating input and output variables as such for every function, as well as by replacing function calls with the according input and output variables following the notation scheme of CBMC-GC. Loops and their data dependencies are analyzed with the help of the PIPS framework. PIPS is capable of exporting loop bodies into a function, which allows to use the same function-wise decomposition described above. Moreover, information about loop ranges is exported for a later use.

Optimization Engine, Translator and Gate Level Optimizer In our implementation all three components run as a single process, which allows to keep multiple sub-circuits in memory at the same time. This approach allows fast switching between the optimization of multiple sub-circuits. The translation phase itself is equivalent to the compilation chain described in Sect. 3.3. Every C assignment is translated into a Boolean circuit preserving its semantics. Moreover, a mapping from input and output variables to wires in the circuit is exported.

Optimization Strategy We propose a heuristic strategy for the optimization engine. The optimization budget is the user specified upper bound on the available computational time for optimization t_{total}. For the initial optimization, a fraction λ_{init} of the total budget is distributed onto all sub-circuits proportionally in the number non-linear gates. Thus, a sub-circuit with twice the number of non-linear gates is assigned twice the initial budget. Loop bodies are weighted by the number of their iterations, functions are weighted by the number of their occurrences. If possible, every sub-circuit is completing at least a single constant propagation and pattern rewriting optimization pass. After the initial optimization, the minimized sub-circuits are ordered by the number of constant inputs, which have been revealed in the previous pass as well as the absolute improvement in the number of non-linear gates in the previous pass. Similarly to the initial pass, loop bodies and functions bodies are weighted by their number of iterations. Based on this order, the most promising sub-circuit is selected next for minimization. After every subsequent minimization pass, the sub-circuit is returned to the queue of circuits to be optimized according the aforementioned ordering criteria. The optimization stops once t_{total} has be reached, or no further improvements are observed.

Composer We implemented a composer that is capable of exporting a single global circuit or a compressed, interpretable representation of the recomposed circuit.

Limitations Currently, the approach described above is unable to decompose recursive functions. However, recursive functions can still be compiled, yet with the standard holistic optimization approach and thus, (in-)efficiency.

7.3 Experimental Evaluation

We first give a description of our experimental setup and the selected functionalities before presenting a comparison with other existing compilers.

7.3.1 Description of Experiments

All functionalities described in the next paragraphs are compiled on a commodity laptop with 8 GB RAM using a single core. According to the compiler analysis by Mood et al. [61], the Frigate [61] and Obliv-C [77] compiler create circuits with the least number of non-linear gates and are therefore the most promising candidates for comparison with the here presented compiler extension. Even though all compilers use different input languages, we ensure a fair comparison as in previous chapters by implementing the functionalities using the same code structure (i.e., functions, loops), data types and bit-widths.

The maximum optimization budget for our extension is set to $t_{total} = 100$ s with the initial optimization run being assigned $\lambda_{init} = 10\%$ of that time in all experiments. For the experiments, we again used the benchmarking applications (see Sect. 2.3) with the following larger parameters:

- *Biometric matching (BioMatch).* For the experiments, we fix the dimension of a sample to $d = 4$ and distinguish a database size of $n = 512$ and $n = 10,000$ samples. The squared Euclidean distance is used as distance function.
- *Matrix multiplication (MMul).* We evaluate a 16×16 matrix multiplication over 32 bit signed integers, fixed-point and floating point numbers.
- *Location-aware scheduling (Scheduling).* We benchmark the functionality for 280 time slots with two 16 bit integer coordinates each. Moreover, we distinguish between the squared Euclidean (*SQ*), as well as the Manhattan distance (*MH*).

Table 7.1 Compiler comparison. Comparison of circuit sizes in the number of non-linear gates for applications compiled by the Frigate, OblivC and extended CBMC-GC compiler. Marked in bold face are significant improvements

Application	Frigate	OblivC	This work	Improvement [%]
BioMatch 512	2,2 M	2,2 M	1,6 M	**27.3**
BioMatch 10,000	43,8 M	43,8 M	31,8 M	**27.3**
MMul 16×16, int	4,2 M	4,2 M	4,2 M	0
MMul 16×16, fix	10.3 M	10.3 M	6.0 M	**41.7**
MMul 16×16, float	89.0 M	80 M	22.8 M	**71.5**
Scheduling 280, Euclidean	666,723	905,487	565,320	**15.1**
Scheduling 280, Manhattan	395,683	385,247	294,000	**23.7**

7.3.2 Compilation Results

Table 7.1 shows the resulting circuit sizes in the number of non-linear gates for all described functionalities and compilers. Furthermore, the improvement over the best of the two results from Frigate or Obliv-C is given. Significant improvements are marked in bold face. Circuit sizes above one million (M) gates are rounded to the nearest 100,000.

For the MMul application over fixed-point and floating-point numbers, we observe that our compiler extension generates circuits that are up to 4 times smaller than the circuits generated by Frigate and Obliv-C. This is due to the more advanced gate level optimizations, which are required to compile the bit-oriented source code efficiently. Yet, for MMul over integers, which barely provides opportunities for optimizations, as only building blocks have to be instantiated, we see the same result between all compilers.

The BioMatch and the scheduling application are more complex and hence, allow various optimizations. For these applications, we observe a size improvement of up to 27%. Even though not shown here, we note that the decomposition approach achieves significant scalability improvements for CBMC-GC. For example, the optimization of the BioMatch application takes roughly the same time for a database size of 16 or 10^7 samples, whereas without decomposition and source code guided optimization, an optimization time of multiple hours is required to achieve the same circuit size, as the circuit in this example consists of more than 1 billion gates. In summary, we observe that the extension to CBMC-GC is capable of creating circuits that are significantly smaller than those created by the previously best known compilers within a reasonable compile and optimization time.

Appendix A
CBMC-GC Manual

CBMC-GC is released as open source and available for download at www.seceng.de/research/software/cbmc-gc/. Older versions with a detailed manual can be found at www.forsyte.at/software/cbmc-gc/. In the following paragraphs, we give an overview of its features and how to use them.

Disclaimer: CBMC-GC is a tool for research purposes and should not be used in a productive environment. We do not guarantee correctness.

Features

At the time of writing, the open-source release of CBMC-GC contains the following features:

- A general compilation chain from ANSI C to circuits (Chap. 3).
- An optimization routine to compile size-minimal circuits (Chap. 4).
- A compiler extension to compile depth-minimal circuits (Chap. 6).
- An adapter for the ABY framework, which implements Yao's garbled circuits and GMW—https://github.com/encryptogroup/ABY.
- A circuit export functionality into the Bristol, Fairplay's SHDL, or the JustGarble circuit formats.
- A plain text circuit simulator that allows to automatically test circuits against C code (compiled with `gcc`).

The code decomposition tools required for parallelization (Chap. 5) and scalable optimization (Chap. 7) are released separately on www.seceng.de/research/software/cbmc-gc/. Both are in an experimental state, as full automatization has not been achieved in all parts, yet.

N. Büscher, S. Katzenbeisser, *Compilation for Secure Multi-party Computation*, SpringerBriefs in Computer Science, https://doi.org/10.1007/978-3-319-67522-0

Compatibility and Compilation of CBMC-GC

CBMC-GC has been developed and tested on LINUX (Ubuntu 14.04 and newer) and should be compatible to most other recent LINUX distributions. CBMC-GC is written platform independently, as far as possible, and hence should be portable to other platforms. For installation, the download has to be unzipped and compiled with make as follows:

```
make minisat2-download
make
```

Compiling Circuits with CBMC-GC

To compile a function into circuit, CBMC-GC has to be invoked as follows:

```
./bin/cbmc-gc application.c --function millionaire
```

If no function name is specified, CBMC-GC will look for a function named mpc_main(). Note, that the return value of this function is automatically interpreted as an output variable. Thus, the millionaires' functionality can be implemented as follows:

```
1 unsigned mpc_main(unsigned INPUT_A, unsigned INPUT_B)
2 {
3   return INPUT_A > INPUT_B;
4 }
```

To successfully compile a functionality, it has to follow CBMC-GC's I/O variable notation, as shown above (see also Sect. 3.3). Further outputs can be specified using the variable name prefix "OUTPUT". CBMC-GC can handle I/O arrays, yet, according the ANSI C standard these cannot be declared as function parameters. Instead, they can be wrapped in a struct or declared in the function's body. Moreover, for successful compilation, a program either has to be bound or a global upper bound has to be set with:

```
--unwind k
```

To limit the circuit minimization phase use (this is essential for large programs), the compilation flag

```
--minimization-time-limit X
```

can be used, where X is an integer value in seconds. To compile circuits with minimal depth the following flag has to be added to the compile command:

```
--low-depth
```

Further documentation of CBMC-GC and a set of application examples can be found in the download.

References

1. Asharov, G., Lindell, Y., Schneider, T., Zohner, M.: More efficient oblivious transfer and extensions for faster secure computation. In: Sadeghi, A.R., Gligor, V.D., Yung, M. (eds.) ACM CCS 13: 20th Conference on Computer and Communications Security, pp. 535–548. ACM Press, New York (2013)
2. Barni, M., Bernaschi, M., Lazzeretti, R., Pignata, T., Sabellico, A.: Parallel implementation of GC-based MPC protocols in the semi-honest setting. In: Data Privacy Management and Autonomous Spontaneous Security - 8th International Workshop, DPM 2013, and 6th International Workshop, SETOP 2013, Egham, September 12–13, 2013, Revised Selected Papers, pp. 66–82 (2013)
3. Beaver, D.: Efficient multiparty protocols using circuit randomization. In: Feigenbaum, J. (ed.) Advances in Cryptology – CRYPTO'91. Lecture Notes in Computer Science, vol. 576, pp. 420–432. Springer, Heidelberg (1992)
4. Beaver, D., Micali, S., Rogaway, P.: The round complexity of secure protocols (extended abstract). In: 22nd Annual ACM Symposium on Theory of Computing, pp. 503–513. ACM Press, New York (1990)
5. Bellare, M., Hoang, V.T., Keelveedhi, S., Rogaway, P.: Efficient garbling from a fixed-key blockcipher. In: 2013 IEEE Symposium on Security and Privacy, pp. 478–492. IEEE Computer Society Press, New York (2013)
6. Ben-Or, M., Goldwasser, S., Wigderson, A.: Completeness theorems for non-cryptographic fault-tolerant distributed computation (extended abstract). In: 20th Annual ACM Symposium on Theory of Computing, pp. 1–10. ACM Press, New York (1988)
7. Berkeley logic synthesis and verification group, ABC: a system for sequential synthesis and verification, release 30916. http://www.eecs.berkeley.edu/~alanmi/abc/
8. Biere, A., Cimatti, A., Clarke, E.M., Zhu, Y.: Symbolic model checking without BDDs. In: Tools and Algorithms for Construction and Analysis of Systems, 5th International Conference, TACAS '99, Held as Part of the European Joint Conferences on the Theory and Practice of Software, ETAPS'99, Amsterdam, March 22–28, 1999, Proceedings, pp. 193–207 (1999)
9. Biere, A., Heule, M., van Maaren, H., Walsh, T. (eds.): Handbook of Satisfiability. Frontiers in Artificial Intelligence and Applications, vol. 185. IOS Press, Amsterdam (2009)
10. Bilogrevic, I., Jadliwala, M., Hubaux, J., Aad, I., Niemi, V.: Privacy-preserving activity scheduling on mobile devices. In: First ACM Conference on Data and Application Security and Privacy, CODASPY 2011, San Antonio, TX, February 21–23, 2011, Proceedings, pp. 261–272 (2011)
11. Bjesse, P., Borälv, A.: Dag-aware circuit compression for formal verification. In: International Conference on Computer-Aided Design ICCAD (2004)

N. Büscher, S. Katzenbeisser, *Compilation for Secure Multi-party Computation*, SpringerBriefs in Computer Science, https://doi.org/10.1007/978-3-319-67522-0

12. Bogdanov, D., Laur, S., Willemson, J.: Sharemind: A framework for fast privacy-preserving computations. In: Jajodia, S., López, J. (eds.) ESORICS 2008: 13th European Symposium on Research in Computer Security. Lecture Notes in Computer Science, vol. 5283, pp. 192–206. Springer, Heidelberg (2008)
13. Bogetoft, P., Christensen, D.L., Damgård, I., Geisler, M., Jakobsen, T., Krøigaard, M., Nielsen, J.D., Nielsen, J.B., Nielsen, K., Pagter, J., Schwartzbach, M.I., Toft, T.: Secure multiparty computation goes live. In: Dingledine, R., Golle, P. (eds.) FC 2009: 13th International Conference on Financial Cryptography and Data Security. Lecture Notes in Computer Science, vol. 5628, pp. 325–343. Springer, Heidelberg (2009)
14. Bondhugula, U., Hartono, A., Ramanujam, J., Sadayappan, P.: A practical automatic polyhedral parallelizer and locality optimizer. In: Proceedings of the ACM SIGPLAN 2008 Conference on Programming Language Design and Implementation, Tucson, AZ, June 7–13, 2008, pp. 101–113 (2008)
15. Buchfuhrer, D., Umans, C.: The complexity of Boolean formula minimization. J. Comput. Syst. Sci. **77**(1), 142–153 (2011)
16. Büscher, N., Katzenbeisser, S.: Faster secure computation through automatic parallelization. In: 24th USENIX Security Symposium, USENIX Security 15, Washington, DC, August 12–14, 2015, pp. 531–546 (2015)
17. Büscher, N., Holzer, A., Weber, A., Katzenbeisser, S.: Compiling low depth circuits for practical secure computation. In: Askoxylakis, I.G., Ioannidis, S., Katsikas, S.K., Meadows, C.A. (eds.) ESORICS 2016: 21st European Symposium on Research in Computer Security, Part II. Lecture Notes in Computer Science, vol. 9879, pp. 80–98. Springer, Heidelberg (2016)
18. Büscher, N., Kretzmer, D., Jindal, A., Katzenbeisser, S.: Scalable secure computation from ANSI-C. In: IEEE International Workshop on Information Forensics and Security, WIFS 2016, Abu Dhabi, December 4–7, 2016, pp. 1–6 (2016)
19. Canet, G., Cuoq, P., Monate, B.: A value analysis for C programs. In: IEEE SCAM (2009)
20. Choi, S.G., Hwang, K.W., Katz, J., Malkin, T., Rubenstein, D.: Secure multi-party computation of Boolean circuits with applications to privacy in on-line marketplaces. In: Dunkelman, O. (ed.) Topics in Cryptology – CT-RSA 2012. Lecture Notes in Computer Science, vol. 7178, pp. 416–432. Springer, Heidelberg (2012)
21. Clarke, E.M., Kroening, D., Lerda, F.: A tool for checking ANSI-C programs. In: Tools and Algorithms for the Construction and Analysis of Systems, 10th International Conference, TACAS 2004, Held as Part of the Joint European Conferences on Theory and Practice of Software, ETAPS 2004, Barcelona, March 29–April 2, 2004, Proceedings, pp. 168–176 (2004)
22. Clarke, E.M., Kroening, D., Yorav, K.: Behavioral consistency of C and Verilog programs using bounded model checking. In: Proceedings of the 40th Design Automation Conference, DAC 2003, Anaheim, CA, June 2–6, 2003, pp. 368–371 (2003)
23. Cuoq, P., Kirchner, F., Kosmatov, N., Prevosto, V., Signoles, J., Yakobowski, B.: Frama-c - A software analysis perspective. In: Software Engineering and Formal Methods - 10th International Conference, SEFM 2012, Thessaloniki, October 1–5, 2012. Proceedings, pp. 233–247 (2012)
24. Cuoq, P., Kirchner, F., Kosmatov, N., Prevosto, V., Signoles, J., Yakobowski, B.: Frama-c - A software analysis perspective. In: SEFM (2012)
25. Dagum, L., Menon, R.: OpenMP an industry standard API for shared-memory programming. IEEE Comput. Sci. Eng. **5**(1), 46–55 (1998)
26. Damgård, I., Pastro, V., Smart, N.P., Zakarias, S.: Multiparty computation from somewhat homomorphic encryption. In: Safavi-Naini, R., Canetti, R. (eds.) Advances in Cryptology – CRYPTO 2012. Lecture Notes in Computer Science, vol. 7417, pp. 643–662. Springer, Heidelberg (2012)
27. Darringer, J.A., Joyner, W.H., Berman, C.L., Trevillyan, L.: Logic synthesis through local transformations. IBM J. Res. Dev. **25**(4), 272–280 (1981)

28. Demmler, D., Dessouky, G., Koushanfar, F., Sadeghi, A.R., Schneider, T., Zeitouni, S.: Automated synthesis of optimized circuits for secure computation. In: Ray, I., Li, N., Kruegel, C. (eds.) ACM CCS 15: 22nd Conference on Computer and Communications Security, pp. 1504–1517. ACM Press, New York (2015)
29. Demmler, D., Schneider, T., Zohner, M.: ABY - A framework for efficient mixed-protocol secure two-party computation. In: ISOC Network and Distributed System Security Symposium – NDSS 2015. The Internet Society, San Diaego (2015)
30. Earle, J.: Latched carry-save adder. IBM Technical Disclosure Bulletin (1965)
31. Erkin, Z., Franz, M., Guajardo, J., Katzenbeisser, S., Lagendijk, I., Toft, T.: Privacy-preserving face recognition. In: Privacy Enhancing Technologies, 9th International Symposium, PETS 2009, Seattle, WA, August 5–7, 2009. Proceedings, pp. 235–253 (2009)
32. Franz, M., Holzer, A., Katzenbeisser, S., Schallhart, C., Veith, H.: CBMC-GC: an ANSI C compiler for secure two-party computations. In: Compiler Construction - 23rd International Conference, CC 2014, Held as Part of the European Joint Conferences on Theory and Practice of Software, ETAPS 2014, Grenoble, April 5–13, 2014. Proceedings, pp. 244–249 (2014)
33. Freedman, M.J., Nissim, K., Pinkas, B.: Efficient private matching and set intersection. In: Cachin, C., Camenisch, J. (eds.) Advances in Cryptology – EUROCRYPT 2004. Lecture Notes in Computer Science, vol. 3027, pp. 1–19. Springer, Heidelberg (2004)
34. Furukawa, J., Lindell, Y., Nof, A., Weinstein, O.: High-throughput secure three-party computation for malicious adversaries and an honest majority. In: Advances in Cryptology - EUROCRYPT 2017 - 36th Annual International Conference on the Theory and Applications of Cryptographic Techniques, Paris, April 30–May 4, 2017, Proceedings, Part II, pp. 225–255 (2017)
35. Ganai, M.K., Gupta, A., Ashar, P.: DiVer: Sat-based model checking platform for verifying large scale systems. In: Tools and Algorithms for the Construction and Analysis of Systems, 11th International Conference, TACAS 2005, Held as Part of the Joint European Conferences on Theory and Practice of Software, ETAPS 2005, Edinburgh, April 4–8, 2005, Proceedings, pp. 575–580 (2005)
36. Goldberg, D.: What every computer scientist should know about floating-point arithmetic. ACM Comput. Surv. **23**(3), 413 (1991)
37. Goldreich, O., Micali, S., Wigderson, A.: How to play any mental game or A completeness theorem for protocols with honest majority. In: Aho, A. (ed.) 19th Annual ACM Symposium on Theory of Computing, pp. 218–229. ACM Press, New York (1987)
38. Goldreich, O., Ostrovsky, R.: Software protection and simulation on oblivious rams. J. ACM **43**(3), 431–473 (1996)
39. Harris, D.: A taxonomy of parallel prefix networks. In: IEEE ASILOMAR (2003)
40. Henecka, W., Kögl, S., Sadeghi, A.R., Schneider, T., Wehrenberg, I.: TASTY: tool for automating secure two-party computations. In: Al-Shaer, E., A.D. Keromytis, V. Shmatikov (eds.) ACM CCS 10: 17th Conference on Computer and Communications Security, pp. 451–462. ACM Press, New York (2010)
41. Henecka, W., Schneider, T.: Faster secure two-party computation with less memory. In: Chen, K., Xie, Q., Qiu, W., Li, N., Tzeng , W.G. (eds.) ASIACCS 13: 8th ACM Symposium on Information, Computer and Communications Security, pp. 437–446. ACM Press, New York (2013)
42. Holzer, A., Franz, M., Katzenbeisser, S., Veith, H.: Secure two-party computations in ANSI C. In: T. Yu, G. Danezis, V.D. Gligor (eds.) ACM CCS 12: 19th Conference on Computer and Communications Security, pp. 772–783. ACM Press, New York (2012)
43. Huang, Y., Evans, D., Katz, J., Malka, L.: Faster secure two-party computation using garbled circuits. In: 20th USENIX Security Symposium, San Francisco, CA, August 8–12, 2011, Proceedings (2011)
44. Husted, N., Myers, S., Shelat, A., Grubbs, P.: GPU and CPU parallelization of honest-but-curious secure two-party computation. In: Annual Computer Security Applications Conference, ACSAC '13, New Orleans, LA, December 9–13, 2013, pp. 169–178 (2013)

45. Irigoin, F., Jouvelot, P., Triolet, R.: Semantical interprocedural parallelization: an overview of the PIPS project. In: ICS (1991)
46. Ishai, Y., Kilian, J., Nissim, K., Petrank, E.: Extending oblivious transfers efficiently. In: Boneh, D. (ed.) Advances in Cryptology – CRYPTO 2003. Lecture Notes in Computer Science, vol. 2729, pp. 145–161. Springer, Heidelberg (2003)
47. Keller, M., Orsini, E., Scholl, P.: Actively secure OT extension with optimal overhead. In: Gennaro, R., Robshaw, M.J.B. (eds.) Advances in Cryptology – CRYPTO 2015, Part I. Lecture Notes in Computer Science, vol. 9215, pp. 724–741. Springer, Heidelberg (2015)
48. Kerschbaum, F., Schneider, T., Schröpfer, A.: Automatic protocol selection in secure two-party computations. In: Boureanu, I., Owesarski, P., Vaudenay, S. (eds.) ACNS 14: 12th International Conference on Applied Cryptography and Network Security. Lecture Notes in Computer Science, vol. 8479, pp. 566–584. Springer, Heidelberg (2014)
49. Kolesnikov, V., Schneider, T.: Improved garbled circuit: Free XOR gates and applications. In: L. Aceto, I. Damgård, L.A. Goldberg, M.M. Halldórsson, A. Ingólfsdóttir, I. Walukiewicz (eds.) ICALP 2008: 35th International Colloquium on Automata, Languages and Programming, Part II. Lecture Notes in Computer Science, vol. 5126, pp. 486–498. Springer, Heidelberg (2008)
50. Kolesnikov, V., Sadeghi, A.R., Schneider, T.: Improved garbled circuit building blocks and applications to auctions and computing minima. In: Garay, J.A., Miyaji, A., Otsuka, A. (eds.) CANS 09: 8th International Conference on Cryptology and Network Security. Lecture Notes in Computer Science, vol. 5888, pp. 1–20. Springer, Heidelberg (2009)
51. Kreuter, B., Shelat, A., Shen, C.: Billion-gate secure computation with malicious adversaries. In: Proceedings of the 21th USENIX Security Symposium, Bellevue, WA, August 8–10, 2012, pp. 285–300 (2012)
52. Kreuter, B., Shelat, A., Mood, B., Butler, K.R.B.: PCF: a portable circuit format for scalable two-party secure computation. In: Proceedings of the 22th USENIX Security Symposium, Washington, DC, August 14–16, 2013, pp. 321–336 (2013)
53. Kuehlmann, A.: Dynamic transition relation simplification for bounded property checking. In: 2004 International Conference on Computer-Aided Design, ICCAD 2004, San Jose, CA, November 7–11, 2004, pp. 50–57 (2004)
54. Lindell, Y., Pinkas, B.: A proof of security of Yao's protocol for two-party computation. Journal of Cryptology **22**(2), 161–188 (2009)
55. Liu, C., Huang, Y., Shi, E., Katz, J., Hicks, M.W.: Automating efficient RAM-model secure computation. In: 2014 IEEE Symposium on Security and Privacy, pp. 623–638. IEEE Computer Society Press, New York (2014)
56. Liu, C., Wang, X.S., Nayak, K., Huang, Y., Shi, E.: ObliVM: A programming framework for secure computation. In: 2015 IEEE Symposium on Security and Privacy, pp. 359–376. IEEE Computer Society Press, New York (2015)
57. Malkhi, D., Nisan, N., Pinkas, B., Sella, Y.: Fairplay - secure two-party computation system. In: Proceedings of the 13th USENIX Security Symposium, August 9–13, 2004, San Diego, CA, pp. 287–302 (2004)
58. McCreary, C., Gill, H.: Efficient exploitation of concurrency using graph decomposition. In: Proceedings of the 1990 International Conference on Parallel Processing, Urbana-Champaign, IL, August 1990. Volume 2: Software, pp. 199–203 (1990)
59. Mishchenko, A., Chatterjee, S., Brayton, R.K.: Dag-aware AIG rewriting a fresh look at combinational logic synthesis. In: Design Automation Conference, DAC (2006)
60. Mishchenko, A., Chatterjee, S., Brayton, R.K., Eén, N.: Improvements to combinational equivalence checking. In: 2006 International Conference on Computer-Aided Design, ICCAD 2006, San Jose, CA, November 5–9, 2006, pp. 836–843 (2006)
61. Mood, B., Gupta, D., Carter, H., Butler, K.R.B., Traynor, P.: Frigate: A validated, extensible, and efficient compiler and interpreter for secure computation. In: IEEE European Symposium on Security and Privacy, EuroS&P 2016, Saarbrücken, March 21–24, 2016, pp. 112–127 (2016)

62. Mood, B., Letaw, L., Butler, K.: Memory-efficient garbled circuit generation for mobile devices. In: Keromytis, A.D. (ed.) FC 2012: 16th International Conference on Financial Cryptography and Data Security. Lecture Notes in Computer Science, vol. 7397, pp. 254–268. Springer, Heidelberg (2012)
63. Muchnick, S.S.: Advanced Compiler Design and Implementation. Morgan Kaufmann, San Francisco (1997)
64. Naor, M., Pinkas, B., Sumner, R.: Privacy preserving auctions and mechanism design. In: EC, pp. 129–139 (1999)
65. Nayak, K., Wang, X.S., Ioannidis, S., Weinsberg, U., Taft, N., Shi, E.: GraphSC: Parallel secure computation made easy. In: 2015 IEEE Symposium on Security and Privacy, pp. 377–394. IEEE Computer Society Press, New York (2015)
66. Nielsen, J.B., Nordholt, P.S., Orlandi, C., Burra, S.S.: A new approach to practical active-secure two-party computation. In: Safavi-Naini, R., Canetti, R. (eds.) Advances in Cryptology – CRYPTO 2012. Lecture Notes in Computer Science, vol. 7417, pp. 681–700. Springer, Heidelberg (2012)
67. Pinkas, B., Schneider, T., Smart, N.P., Williams, S.C.: Secure two-party computation is practical. In: Matsui, M. (ed.) Advances in Cryptology – ASIACRYPT 2009. Lecture Notes in Computer Science, vol. 5912, pp. 250–267. Springer, Heidelberg (2009)
68. Pouchet, L.N.: Polyhedral Compiler Collection (PoCC) (2012)
69. Pullonen, P., Siim, S.: Combining secret sharing and garbled circuits for efficient private IEEE 754 floating-point computations. In: Financial Cryptography and Data Security - FC 2015 International Workshops, BITCOIN, WAHC, and Wearable, San Juan, Puerto Rico, January 30, 2015, Revised Selected Papers, pp. 172–183 (2015)
70. Robertson, J.E.: A new class of digital division methods. IRE Trans. Electron. Comput. (3), 218–222 (1958)
71. Schneider, T., Zohner, M.: GMW vs. Yao? Efficient secure two-party computation with low depth circuits. In: Sadeghi, A.R. (ed.) FC 2013: 17th International Conference on Financial Cryptography and Data Security. Lecture Notes in Computer Science, vol. 7859, pp. 275–292. Springer, Heidelberg (2013)
72. Schröpfer, A., Kerschbaum, F., Müller, G.: L1 - an intermediate language for mixed-protocol secure computation. In: Proceedings of the 35th Annual IEEE International Computer Software and Applications Conference, COMPSAC 2011, Munich, 18–22 July 2011, pp. 298–307 (2011)
73. Songhori, E.M., Hussain, S.U., Sadeghi, A.R., Schneider, T., Koushanfar, F.: TinyGarble: Highly compressed and scalable sequential garbled circuits. In: 2015 IEEE Symposium on Security and Privacy, pp. 411–428. IEEE Computer Society Press, New York (2015)
74. Wallace, C.S.: A suggestion for a fast multiplier. IEEE Trans. Electron. Comput. **EC-13**(1), 14–17 (1964)
75. Yao, A.C.C.: Protocols for secure computations (extended abstract). In: 23rd Annual Symposium on Foundations of Computer Science, pp. 160–164. IEEE Computer Society Press (1982)
76. Yao, A.C.C.: How to generate and exchange secrets (extended abstract). In: 27th Annual Symposium on Foundations of Computer Science, pp. 162–167. IEEE Computer Society Press, New York (1986)
77. Zahur, S., Evans, D.: Obliv-c: A language for extensible data-oblivious computation. IACR Cryptology ePrint Archive **2015**, 1153 (2015)
78. Zahur, S., Rosulek, M., Evans, D.: Two halves make a whole - reducing data transfer in garbled circuits using half gates. In: Oswald, E., Fischlin, M. (eds.) Advances in Cryptology – EUROCRYPT 2015, Part II. Lecture Notes in Computer Science, vol. 9057, pp. 220–250. Springer, Heidelberg (2015)